A Contemporary Theory of Mathematics Education Research

Tony Brown

A Contemporary Theory of Mathematics Education Research

Springer

Tony Brown
Manchester Metropolitan University
Manchester, UK

ISBN 978-3-030-55099-8 ISBN 978-3-030-55100-1 (eBook)
https://doi.org/10.1007/978-3-030-55100-1

This Springer imprint is published by the registered company Springer Nature Switzerland AG
The registered company address is: Gewerbestrasse 11, 6330 Cham, Switzerland

Preface

This book addresses the domain, purpose and functioning of research in mathematics education. What is mathematics teaching? How do we improve mathematics teaching? Why do we want to improve mathematics teaching? What do we understand by improvement? Mathematics education research addresses many such questions. And although research and the scientific theories it produces may never reveal the truth, theories have proved very useful in guiding us around an infinite landscape, even if ultimately each scientific model or theory reveals the limits of its own functionality, style or endurance. But more generally, any story we tell about what mathematics is, or what a learner is, or what we are trying to do, will eventually become out of kilter with the times. New demands on "people" and new demands on "mathematics" change what both of them are and how they respond to each other. Research has as much to do with working out where we could go as it is with assessing where we are now. The current state of affairs can be depicted in many ways, where alternative mappings of our pasts and presents open alternative trajectories into the future. This provokes a more general question that motivates this book: What can theories do? How do we situate our theories in relation to other theories, past and present? Do new theories replace old ones or sit alongside them? Many contributions to theory in mathematics education comprise individual journal articles or chapters in edited collections.[1] Theories depend on the questions that we ask and the world views that we presuppose and typically theories, especially social theories are time-dependent, and need persistent updating. So in presenting "a contemporary theory of mathematics education research" the intention is to unsettle some of the common presumptions of mathematics education research in generating new ways of looking rather than to suppose any final resolution might be reached. But in unsettling, the hope is that new ways of understanding the interface between humans and mathematics will be suggested and stimulate life thereafter.

---

[1] As examples, Springer has published theory-oriented chapters by over 100 authors in the following edited collections alone: Sriraman and English (2010); Bikner-Ahsbahs et al. (2016); and Ernest (2019b).

How then do we understand renewal in mathematics education practice? When I started my career some 40 years ago, student-centred approaches to mathematics, including problem solving, investigations and project work, seemed to be on the ascendance in curriculum reform agenda in parts of England, certainly in London where I was working as a secondary school teacher at the time. School mathematics had often received a bad press in terms of children's attitudes towards the subject and many adults still claim unfortunate experiences in their own schooling. Teachers at the time were keen to find ways of making mathematics intrinsically more interesting for a wider body of students. The quest to centre education on personal development underpinned my aspirations in entering teaching where education was to be seen primarily in terms of valuing each child as an individual rather than merely meeting required standards across a given group. Many London schools were following an individualised learning scheme fostering such aspirations. Nationally, the Association of Teachers of Mathematics also pursued more investigational approaches. This move to new models of practice in England paralleled and possibly pre-dated the so-called "math wars" in the United States between traditional methods and the constructivist philosophies that ultimately had more international traction in both practice and research. A sustained British government backlash throughout the 1990s, however, resulted in prescribed national curriculums for both teachers and students in England in which student-centred approaches became more tightly structured around legal specifications of curriculum content and preferred but "non-statutory" methods of delivery. Under a banner of "back to basics" successive policies initiated much closer definition and scrutiny of classroom practice towards achieving wider inclusion. Reasons cited for this backlash included right wing politicians claiming that given difficulties with teacher supply, keenly felt in secondary mathematics at the time, the average teacher could not teach to such high-minded ideals. It argued that there was no point having a preferred form of teaching if teachers could not be trained to teach in this way, better to tell the teachers what to do in line with a centralised definition of expectations, rather than let left wing academics in teacher education colleges lead them astray into so-called "progressive" methods. Teacher education in England thus became more prescriptive and school-based with increased reference to the new curriculum frameworks, perhaps at the expense of aspirations for more autonomous teacher functionality. Schooling around the world, meanwhile, became increasingly shaped and judged by its perceived capacity to deliver success in terms of international competitiveness linked to economic agenda, often as indicated through performance in comparative tests. The shift of policy seemed to be reflected in England's rankings. It moved upwards from 18th to seventh position on the skills-focused TIMSS in 2007, whilst dropping from eighth to 25th on the (possibly) more problem focused PISA in 2006. Hooray, except the government then complained about the dip in PISA that had now become the more prominent instrument. Priorities do not always pull in the same direction and often change. In mathematics education, choices do sometimes need to be made between promoting exam success, supporting future professional functionality, enjoyment for the subject, advanced mathematical behaviour, inclusion for a broad range of pupils, etc. and there are differing ways of promoting and measuring each of those alternative priorities. Research in the field increasingly finds its terms of reference set according to assessment driven requirements and researchers have become complicit in promoting particular conceptions of teaching and in constructing the field as an ideological battleground, for example, in commissioned research where briefs can presuppose improvements of some kind. Such complicity, combined with the relative insularity of the field, has deflected many mathematics education researchers from investigating other world visions that might define us and serve us in different ways.

There is, however, need to pause to consider how we are drawn into our diverse motivations. For example, is high performance according to a scientific measure like PISA or TIMSS necessarily a good thing for a country? Insofar as such comparative instruments aspire to a standardisation of school mathematical priorities there is a risk that countries are served differently, and not necessarily according to their specific needs. A casual glance at TIMSS test items reveals a very specific version of mathematics, centred on basic skills, short closed questions, in bland "real life" situations. The format is echoed in school tests and a host of materials widely available to parents designed to prepare children for such tests. But to represent mathematics as universal, spanning nations and generations, in such a singular fashion comes at a price. The resultant conceptions of school mathematics now define and police everyday practice. At a major mathematics education international conference a Mexican delegate spoke of how the exercises made her country subservient to American priorities for school mathematics. An Ethiopian educator depicted a situation in which teachers and students were obliged to engage with pedagogical formations largely unrecognisable in his country situation. Meanwhile, a Finnish commentator indicated that her country's high performance still required re-evaluation of their national practices in terms of the newly dominant international discourse and its stated priorities. But cutting across those sorts of issues we may ask if widespread success in such measures has any bearing in a country's ability to produce top-level mathematicians. Or conversely, if the aspirations of TIMSS are so bland, is wider inclusiveness necessarily a good indicator of wider basic functionality in the subject in any useful way? Many mathematics education researchers would concur on the limitations of these comparative instruments, but still they remain recurrent points of reference in so many reports on mathematics education research, including my own, including this one, as an attempt to reach out to mutually recognisable themes. And my use of such consensual issues weakens my individual voice in its attempted compromise in the name of a short cut to communication. We know that these shared points of reference are limited but we still carry on using them and allow them to orientate, format and exchange our evaluative efforts even though we secretly acknowledge their wobbly foundations. This need to chip away at our own false premises is a key ingredient of this book's discussion. But sometimes these false or alternative premises are imposed on us through official agencies more concerned with wider policy-driven social management than with more precise research-led ideals. In short, mathematics in schools is governed by ideologies that have varying shelf-lives, domains of relevance and underlying motivations.

A reviewer of the proposal for this book was keen to capture the "take away" message that lay beyond the analysis I had presented, and I am similarly keen to establish this message from the outset. Here goes: People and mathematics are in persistent co-evolution and any account of their mutual interaction requires a flexibility of language, where the operation of that very flexibility is often the instrument of change. That is, our understanding of what it is to be a person is persistently changing, as is our understanding of what we want mathematics to do, but where certain aspects of mathematics, unlike other disciplines, remain remarkably stable. Research must comprise the analysis of these understandings targeted at developing actions through which the mutual evolution can be better understood and activated according to newly defined priorities. Similarly, teacher education is presented as a challenge for student teachers to research their own process of becoming a teacher through critically analysing their own engagement with mathematics and their early attempts to teach it. We all need to adopt a critical attitude towards our past assumptions or contemporary officialdom that constrain our thinking into specific pathways. In this book, the attempt is to see research as the motor with which to achieve this. Mathematics is not just out there waiting to be found – the very content of mathematics is a function of human processing that is necessarily governed by historical processes, human priorities and power relations, but where these processes, priorities and relations will be persistently in motion and potential conflict.

The reviewer also asked, in recognising my normal home base of social theory: "is it possible to discuss the implications about classroom instruction based on this work? That is, is it possible to base on this work to at least discuss about the way to teach mathematics in classroom for maximizing students' learning opportunities?" I do propose to do that as I have in earlier work. This is a reasonable request. Lots of people spend lots of their time in their formative years in mathematics classrooms and it is an obvious forum in which to consider the issues being addressed in this book. I have discussed classroom work with children in my earlier work (e.g. Brown 2001). In this present book, my main point of intervention is with student teachers. I see this challenge in terms of how teachers might re-think their participation in their lessons tomorrow rather than assuming that structural changes are necessary before one can begin new forms of practice. We cannot await perfectly prepared children, in a perfect classroom, with a perfect curriculum, in a perfect future. It is possible to rethink teacher/student/mathematical relationships in any current setting. The main instrument proposed comprises student teachers carrying out analysis of their own emerging teaching practice and in relation to their own attempts at mathematical tasks. Yet the book also discusses the many curriculum and institutional constraints that operate on classroom practice in mathematics.

I will loosen any assumption that we are *only* in the business of supporting classes of students. "Classes" are very much a time-dependent educational construct, normal, perhaps, for most young people presently. Yet, this has been the case for much less than a century, only since contemporary social organisation has required such arrangements for a lot of young people, where they experience their mathematical learning as a collective experience. The reviewer's concern for "maximising students' learning opportunities" will always be according to a particular agenda where consensus, even among researchers, is typically evasive. Priorities do not always pull in the same direction. We must not remain trapped in old models as a stopgap simply because we have not learned to replace them. In this respect, I differ from the reviewer's preferences in that I want to step back and spend more time on deciding how we reach the metrics by which we decide gradations. I am on the side of trying to explore alternative productive ways of looking, understanding how theory might work, rather than moving too quickly to decisions about preferred courses of action. I am more concerned with how a social scientist would make sense of mathematics education research than I am with deciding what teachers should do next. There's enough of the latter already. But at the same time I am talking to the mathematics educators that I have encountered over the years rather than to the theorists who would disown my simplistic use of their work. The book attempts to work more generally at the human/mathematics interface in terms of more widespread participation but where the terms of that participation are left open as part of the pedagogic encounter.

So often in my reading of mathematics education research, conclusions have pushed for more active participation by students in mathematics rather than mere compliance with contemporary norms. Yet the apparent reality in schools has often been towards ever more corporate models of practice governed by a competitive neo-liberal ethos, where structural priorities trump autonomous action by teachers, preferred pedagogical routes or test performance trumps mathematical exploration by students. Research typically has very little impact on conceptions of policy and a very weak or indirect impact on actual practice. The voice of the policy maker saying this *must* be done will often be stronger than that of a mathematics education researcher suggesting productive courses of action. Also, outlets for mathematics education research are usually insistent that the mathematical elements of any research are pinpointed within any wider depiction of the educational context. But the negotiation of this wider context by researchers, teachers and children inevitably shapes the mathematics that is encountered. As the author of this book, I am also very conscious that my own personal perspectives have evolved through that of being school pupil, university mathematics student, trainee teacher, schoolteacher, teacher educator, researcher, professor and author. My personal assessments of research and its relevance are a function of the stage I have reached in my career and affect my views more than actual changes in historical circumstances. Perhaps young teachers typically aspire to more emancipatory approaches with their individual classes of children, whilst mid-career teacher educators and researchers are more attentive to models of practice that can be shared across communities. The insertion of one's own delusional personal history into an account of supposed wider trends always requires the pursuit of multiple perspectives built through successive new demands and frequently changed minds. This book comprises my attempt to speak from the present, to take the chance on asserting a new theory acknowledging the essential collectivism built into any point of view, as seen from my personal pathway as I currently understand it.

A Singular Journey into Mathematics Education

The remainder of this preface will set out some of the moments that have resulted in the perspectives presented above and in the chapters to follow. The material for this, my tenth book, is drawn from my work in mathematics education produced since the 2011 publication of two books, *Becoming a mathematics teacher*, a write up of two empirically based primary mathematics teacher education projects funded by the UK Economic and Social Research Council, and *Mathematics education and subjectivity*. The intervening period however has also been devoted to a teacher education project that I led culminating in two books without a specific mathematical theme entitled *Teacher education in England* (Brown 2018) and *Research on becoming an English teacher* (Brown et al. 2019). This current book will be an attempt to make cumulative sense of my complete body of work in mathematics over my career, an attempt at an articulation of a unifying theme, a composite argument, or even a "take away message". The autobiographical dimension to this preface situates the current work into a longer-term professional trajectory with view to offering some explanation of how I have ended up where I am today and why I think some of the things that I do.

I was fortunate in my early career to encounter two influential figures in the formation of Britain's Association of Teachers of Mathematics, Dick Tahta and Bill Brookes, who were responsible for my relatively counter-culture engagement with mathematics education research from the outset. Following my first degree in mathematics and economics, I studied secondary mathematics education, with Dick in Exeter (1978–79) for my initial teacher education, and later in my PhD with Bill in Southampton (1985–87). I was also fortunate that my first two jobs also provided havens of "progressive" experimentation; 3 years teaching secondary level mathematics (11–19-year-olds) at Holland Park School in central London, followed by a similar period as a primary level teacher educator with the organisation Voluntary Services Overseas in the tiny Caribbean island of Dominica.

Freed from the regulative structures that shape early teacher practice in many countries today, I found myself asking questions that do not quite fit with the way in which school education is often approached in the current climate. It was Dick who was rather troubled by the idea that teaching had become characterised as providing explanations to prescribed questions. He preferred rather that teaching would be about the displacement of perspectives with the teacher responding to a child's question by providing another question in return. I have pursued this approach doggedly over the years to the irritation of my students, many of whom have given up any hope of receiving a straight answer from me to their questions. Further, Dick argued that children were rarely wrong in what they said, they were simply addressing a different question, rather than the one supposed to be in the teacher's alien language.

I was left to work things out for myself with the occasional book nudged my way but with an accompanying expectation that I found my own books as well. For example, Dick passed me an essay by Roland Barthes called "Writers, teachers and intellectuals", a paper not included on many training year reading lists, even in the heady late 1970s. In this article, written shortly after the student uprisings in Paris in 1968, I encountered my first reference to Jacques Lacan, a psychoanalytical practitioner and theorist influenced by Sigmund Freud and the philosopher Georg Wilhelm Friedrich Hegel. Lacan was to later become my main intellectual influence for over 30 years. This iconic quote from Barthes' essay appeared in the final chapters of my training year dissertation, in my PhD, in my first book and in several other places since, including in a tribute I read at Dick's funeral.

> Just as psychoanalysis, with the work of Lacan, is in the process of extending the Freudian topic into a topology of the subject… so likewise we need to substitute for the magisterial space (the word delivered by the master from the pulpit above with the audience below, the flock, the sheep, the herd) – a less upright, less Euclidean space where no one, neither teacher nor students, would ever be *in his final place*. One would then be able to see that what must be made reversible are not social "roles" (is there any point squabbling for "authority", for the right to speak?) but the regions of speech. Where is speech? In locution? In listening? In the *returns* of the one and the other? The problem is not to abolish the distinction in functions (teacher/student…) but to protect the instability and, as it were, the giddying whirl of positions of speech. In the teaching space, nobody should anywhere be in his place (I am comforted by this constant displacement: Were I to *find my place*, I would not even go on pretending to teach, I would give up). (Barthes 1977, pp. 205–206)

Three years teaching secondary level at Holland Park School in central London followed, where it was true that children often did not stay in their place for very long. This wholly exciting urban environment in the Notting Hill area of London where over 70 nationalities and as many languages were represented in the school defied the production of clear formulae for teacher success. My task as a teacher was to make things happen, and to go with the flow, not to follow a prescribed route, even if that were possible. Often, this was not comfortable, but it could be exhilarating to enable children to choose their own routes through mathematics in an environment that encouraged that. And those years did serve as an anchor to my future understandings of what it is to be a teacher. Inner city schools offer resistance to the best intentioned teaching approaches. There are immediate challenges that do not allow teachers to wait for an ideal state of affairs before proper teaching can begin. Moreover, the school itself went through multiple changes of identity in response to evolving ideas of how educational policies should be shaped and prioritised. London schools, for example, have been more recently successfully responsive to demands for better exam results to the possible demise of earlier more "progressive" approaches tolerated by the city's earlier left-wing administration.

I spent three subsequent years in Dominica, a small rain-forested island of 29 miles by 16, but with several mountains higher than any of those in my native England. I was working with teachers training on the job mainly in remote rural primary schools (5–11-year-olds) some without electricity or running water. The teachers were straight from their own schooling, where many of these teachers had not secured their own 16+ high school success in mathematics. Many of the lessons that I observed entailed a verbose teacher preaching to the flock. My teacher education strategy entailed persuading the teachers to experiment with giving more opportunity for the children themselves to structure their own learning and by talking with each other. For the teacher the challenge was to say fewer words but for those fewer words to be selected more carefully with view to having higher impact. In an island of just 70,000 people, I was given unusual responsibilities for someone in his mid-twenties, including the writing of the national primary school mathematics curriculum, an activity that enabled me to build a grasp of children's early mathematical development, but also to think of ways in which a curriculum could be presented to make problem solving more prominent. More importantly for my later work, it extended my range of interest in mathematics learning to span the whole of childhood through to university level study. On the island, with no TV or internet, at best my daily news comprised 10 minutes of crackles from the BBC World Service radio. I was untroubled by and largely unaware of the popular concerns of the day.

I commenced my PhD centred on these experiences following a meeting with Bill Brookes during a break between my second and third years in the Caribbean. Bill's introductory directions had advised seven or so books including Wittgenstein's (1983) *Philosophical investigations*, Ricoeur's (1981) *Hermeneutics and human sciences*, Polanyi's (1978) *Personal knowledge*, Schütz's (1962) *Problem of social reality*, Collingwood's (1982) *Autobiography* and oddly, Raymond William's (1983) glossary *Keywords*. Each taking up valuable luggage space in advance of a further year's isolation where my sole phone call comprising 5 minutes of conversation with Bill cost me a few days' wages. But there was nothing in that reading with any sort of mathematical theme. If there was any sort of unifying theme it would be to do with how we conceptualise communication and knowledge through the medium of language. It was Bill, following the famous Oxford historian R. G. Collingwood, who alerted me to the idea that explanations are not so much statements of fact but more generally a function of the question that had been asked, or the audience to whom the answer is addressed. The questions that we ask, reveal the perspective that we are taking and the world view associated with that perspective. The questions we answer may well reveal who it is that we are talking to and what we hope to achieve in doing this. That is, our answers reveal the demands to which we are responding. On an island, where the tiny college library had recently been blown away by a devastating hurricane in 1979, my reading for the year was highly focused on these difficult books which defied easy synthesis to someone educated in mathematics and economics. Yet they each provided interesting and challenging ways of thinking about humans interacting in language, here specifically, young Dominican children engaged in experimental mathematics. Beyond my everyday duties as a teacher educator the empirical work focused on a group of seven teachers who I was supervising, seeking to better manage their use of language in lessons featuring investigational mathematics. The subsequent PhD, completed after further fieldwork in London schools, was not an engagement with contemporary research debates but a discussion of how children and teachers shared mathematics as seen through alternative theoretical filters. At my first (unsuccessful) interview for an academic job at London University as my PhD approached completion it was pointed out that an FLM "research" paper that I had submitted to the panel had no reference list (Brown 1987a). The two spells with Dick and Bill had been remarkably devoid of any reference to mathematics education research prevalent at the time. I was also painfully aware that the references list in my PhD thesis did not reach the bottom of the second page with very few mentions of work specifically in the field of mathematics education (Brown 1987b). A reviewer of the current manuscript despaired in a similar way.

I had returned to England at a time when there was a strong preference to recruit teacher educators with recent school teaching experience, where my 3 years in Dominica counted as distance from the classroom rather than relevant experience. For that reason, I returned to school teaching after completing my PhD. I decided to work at a middle school (9–13-year-olds) in the Isle of Wight (23 miles by 13) as a mathematics specialist to retain and develop my connection to both primary and secondary education. Through this period, I applied for many academic jobs without success. Despite my academic and school experience shortfalls, on my fortieth application for a teacher education post, I finally secured my first academic post in Manchester a couple of years later (1988), mainly involved in the education of primary teachers. Manchester has always been a rich environment for intellectual discussion and it was especially exciting for teachers and teacher educators wanting to consider new ways of understanding their work. Weekly meetings of the Teaching and Learning Enquiry Group continued for 9 years and centred on discussions of mathematics teaching practice, with relatively little attention to the wider work of mathematics education research.

My belated temptation to reach out to the mathematics education research community in the early 1990s required a little more awareness of other people's work. At the time, mathematics education research was firmly centred in debates concerning what was called constructivism. To enter serious debate on social aspects of mathematics education research and get published one had to position one's ideas in relation to the discussions taking place under that name. The early days of my more formal research engagement and my first attendance at the conference on the Psychology of Mathematics Education in Lisbon in 1994 were dominated by a debate between radical constructivism, referenced to the individualist developmental psychology of Piaget,[2] and Social Constructivism,[3] which increasingly identified with Vygotsky and perhaps a more sociological perspective. The apparent options for a young researcher like myself at the time was either to take sides in these alternative routes[4] or, given my counter-culture tendencies, to show that both were providing partial perspectives that would surely reach their shelf life in due course. I had misgivings from the start as the American dominated international research landscape, where many researchers pushed for constructivist-oriented "reform" referenced to the *math-wars* centred more on problem solving approaches, seemed not to have noticed the Piagetian child-centred philosophies that had been the norm in English primary schools for a couple of decades. Here, children's minds followed natural paths of development as individuals, where that development determined

[2] e.g. von Glasersfeld (1991, 1995); Steffe and Kieran (1994).

[3] e.g. Cobb and Bowers (1999); Lerman (2000).

[4] Confrey (1991) provided an influential comparison of the two trajectories around that time.

what they could do. But at the time of my earlier teaching in London in the early 1980s, a Marxist tradition had emerged to resist the apparent truths of Piaget's notion of a naturally developing child that had underpinned this child-centred pedagogy. Valerie Walkerdine was a prominent figure in this movement, introducing poststructuralism to the fringes of a British mathematics education research community unaccustomed to such philosophical orientations, nor to existential critiques of Piaget's psychology. Valerie kindly joined the "Teaching and Learning Enquiry Group" for two separate days after we had spent some time reading her book, *The mastery of reason*. The following paragraph from that book gives a flavour of her highly controversial opposition to the styles of child-centred teaching then so prevalent in English primary schools.

> modern scientific accounts, like Piaget's, can be understood as implicated in the production of our modern form of government – the democratic government of reason. Foucault goes beyond the idea of ideologies as relatively autonomous, as sign systems, to discourses which produce a truth, which claim to be an account of "the real"... For me the importance of this work lies in the way in which actual social practices may be discursively regulated by the production of "truths", "knowledges" about children, for example, which claim to tell the truth about child development ... creating a normalising vision of a "natural child". (Walkerdine 1988, p. 5)

The underlying claim here is that there is no such thing as "natural development" – a rather shocking notion for teachers schooled in the work of Piaget. The label masks something altogether more complicated in a world not defined by such clear signposts. Walkerdine was following Lacan's psychoanalytic theory where the words that label things (i.e. the word-thing couple) are rather less secure than is often supposed. The landscape is not defined by a set of agreed things with given labels, or not for very long anyway. For Lacan, the individual's understanding of who she is, is encapsulated in her response to an ever-shifting symbolic network. This symbolic network directs and controls their acts, but without knowing what it wants. The network comprises the discourses that I inhabit, try out for size, explore myself through, in which I see myself reflected, etc., and ultimately learn "who I am" in an infinitely contingent manner. Lacan, writing in the late 1960s, was explicitly critical of his contemporary Piaget: "The Piagetic error ... lies in the notion of what is called the egocentric discourse of the child, defined as the stage at which he lacks what this Alpine psychology calls reciprocity. ... The child, ... does not speak for himself, ... – they don't speak to a particular person, they just speak to nobody in particular" (Lacan 1986, p. 208). Lacan's analysis sees our actions as always responding to some perceived demand in the social network, but that we never fully reconcile the conflicts between the multitude of apparent demands that we encounter.

The Piaget/Vygotsky debate between those of different persuasions was never likely to settle and contemporary protagonists occupy similar sorts of territory, albeit asking a variety of incommensurate questions that defy the achievement of consensus (e.g. Roth 2010). A specific departure from the dominance of the *Psychology of Mathematics Education* conference was the emergence of *Mathematics Education and Society* predicated on a less individualised conception of mathematical development. Rather, social systems made demands on their citizens that shaped them according to conventional expectations, but where radical politics might be able to resist the oppressive dimensions of such expectations. And that bumpy environment provided the backdrop to the emergence of my own ideas.

The re-ascendance of Vygotsky's social theory in the noughties within the banter of mathematics education research derived from the earlier debate described above to become an alternative mainstream in mathematics education research. Debates surrounding this trend provided a common theme within a group that I co-founded with Julian Williams and Yvette Solomon in Manchester in more recent years; Mathematics Education and Contemporary Theory. This conference emerged out of a small group reading Badiou's *Logics of Worlds* over an 18-month period shortly after it was published. Three conferences in Manchester were held over the period 2011–2016 and guest edited the proceedings for two special issues of the journal *Educational Studies in Mathematics* (Brown and Walshaw 2012; Brown et al. 2016). Our community had emerged from earlier collaborations on four edited collections, which included multiple chapters by over 20 authors from the group (Walshaw 2004, 2010; DeFreitas and Nolan 2008; Brown 2008; Black et al. 2009). Primarily through discussion and the advance circulation of papers each of the invitational conferences with about 45 delegates from multiple countries each time asserted, or at least explored, conceptions of theory in the development of mathematics education research. The production of the special issues comprised a key activity associated with the conference, where delegates were invited to submit papers for consideration. The papers in the *Special Issues* sought to explore the frontiers and possible futures of mathematics education research through considering how alternative theoretical lenses enabled new possibilities in apprehending practice. They offered theoretical, narrative, empirical and practical applications of alternative concepts to and around the field of mathematics education to that end, the conferences and *Special Issues* contained essays that made a case for theory. A recurrent issue in processing the papers for publication was the need to insist on an explicitly mathematical dimension in socially oriented discussion.

A common debate centred on how the supposed trajectory from Vygotsky mentioned earlier could be understood. Prominent members of the group Radford, Williams, Solomon and Roth identified with Vygotsky and referenced their work to that trajectory. For example, Roth and Lee (2007) discuss Vygotsky's neglected legacy, Williams (2015) mediated between Bourdieu and neo-Vygotskian perspectives. In contrast, Bibby (2010) argued that Vygotsky's Zone of Proximal Development was not the neutral place one might imagine, an individual might not always respond well to being included in the clan. My own work (e.g. Brown 2016) and that of my colleague Alexandre Pais (2015, 2016) resisted the nodal power given to Vygotsky and Piaget suggesting they had a normalising effect on the discourse of mathematics education research since their work widely underpinned its ideologies and supposed terrain. Meanwhile, Llewellyn (2018) followed Walkerdine's use of Foucault to argue against the normalising effect of mathematics curriculum. Similarly, Nolan (2016) followed Bourdieu in seeing school practices as producing and reproducing "opinions" or notions of "the good mathematics teacher", thereby shaping identity and agency in "becoming" a teacher within institutional structures. De Freitas and Walshaw (2016) have discussed a range of alternative theoretical frameworks in mathematics education research, including Vygotsky, Foucault and Lacan.

My own involvement in the conferences sometimes pursued a discussion centred on my own book *Mathematics education and subjectivity* (Brown 2011) and is continued in this present book. I had initiated some dialogue primarily by writing an ESM article (Brown 2008b) that provided a critical analysis of an ESM special issue on semiotics edited by Saenz Ludlow and Presmeg (2006), where many papers drew on Piagetian and Vygotskian psychological models. Presmeg and Radford (2008) responded to my article in the same journal. My subsequent book provoked lengthy responses in ESM from both Wolff-Michael Roth (2012) and Alexandre Pais (2015, 2016). In turn I was given the opportunity to respond in the same journal to Roth's Vygotskian critique of my work (Brown 2012), further developed in this book as Chapter Eight. It would be inaccurate to say that the debate was resolved but this present book provides my further contribution to an on-going discussion, although I restrict my discussion of Vygotsky in other parts of this book to allow space for my own preferred styles of analysis.

For a long time my principal guide has been Slavoj Žižek, who is absolutely contemporary. I have read more than 40 of his books over a 30-year period but I cannot read them as fast as he writes them. They are all less than two metres from the computer on which I write. I have sought in this current book of mine to make some very limited reference to a major book that he published just 3 months ago. Despite it being my major lockdown project, I have not yet grasped his extensive references to "unorientable" topological spaces like mobius strips, cross-caps and Klein bottles as alternatives to our rectilinear or "statified" obsessions in making sense of temporal flow. I have seen him speak many times and once spent an entire day driving him around the north of England. He autographed the dashboard of my car. I have a Žižek T-shirt. He has provided my main route into Lacan and Hegel, and it was fun. Meanwhile, I have also attended several lectures by his friend, Alain Badiou, formerly chair of Philosophy at the prestigious École normale supérieure in Paris, who also features prominently in this book. I have a definite attraction to listening to major philosophers providing concrete examples from the world that we live in right now even if I can't quite pull off the minimal distance that Žižek advocates in the introduction to the recent book that I have mentioned. And he has published another book since then.

Manchester, UK Tony Brown

Contents

About the Author

Tony Brown is Professor of Mathematics Education at Manchester Metropolitan University. His research mainly considers mathematics education and teacher education through the lens of contemporary social theory. Tony has written nine previous books and many journal articles in these areas. He completed three projects for the Economic and Social Research Council on the theme of Primary Mathematics Teacher Education. Tony co-organised the three conferences on Mathematics Education and Contemporary Theory, held at MMU and co-edited special issues of *Educational Studies in Mathematics* from the material that arose.

Tony has also had a long-standing interest in professionally oriented research, typically carried out by senior professionals working on doctoral studies analysing their own practice. His own students have researched areas as diverse as mathematics education, teacher education, science education, emergency medicine, police training, emotion in special needs education, educational links with industry, English education, race and ethnicity, popular music education, global education in development contexts, early years education, school leadership, and digital media. Seven of these doctoral projects have led to books.

Originally from London, Tony attended the Universities of Kent at Canterbury and Exeter before returning to central London where he taught mathematics for 3 years at Holland Park School. The next 3 years were spent as a Mathematics Teacher Educator for Volunteer Services Overseas in Dominica in the Caribbean. In 1987, he completed his PhD at Southampton University, which focused on language usage in mathematics classrooms, based on data collected in Dominica and London. After a spell as the Mathematics Coordinator in a middle school on the Isle of Wight, Tony moved to Manchester Metropolitan University (then polytechnic) in 1989 to spend some more years in teacher education. He became a Professor at MMU in 2000. During 2003 and 2004, Tony was based at the University of Waikato where he was the first Professor of Mathematics Education in New Zealand. Since then he has been back at MMU writing, teaching and carving out a miserable existence of unsuccessfully applying for research grants.

Chapter 1
Introduction

Mathematical ideas have become very familiar to us as compulsory elements of most people's education but more generally within everyday life. The acquisition of mathematical ideas and processes is often experienced as a formal demand to ensure that the acquisition of certain mathematical ideas in prescribed forms has taken place, with suitable checks and balances to measure this acquisition. But the original desires for including mathematical ideas in our everyday lives appear to have been rewritten to meet specific contemporary caricatures of mathematics and the supposed world that it now serves. In recent years, some of these caricatures have ostensibly been produced to facilitate the obsessive "audit culture" that emerged in the 1990s where everything needed to be measured and compared (Strathern 2000). Consequentially, the formal task of teaching has been increasingly recast as the "delivery" of so many commodities according to preferred metrics. But what is the collateral concealed in the forms of mathematics crafted as commodities in this way? What is delivered, as might arrive in a supermarket delivery van, and what damage is done by the plastic wrapping that we barely noticed until our seas filled with plastic and we feared that it was a potential vector for Covid-19? That is, what is embedded in the materials and practices through which people encounter what is called "mathematics" today? The dominant commodity forms in mathematics education govern both our practices and our analyses of those practices and risk displacing mathematics as a living response to everyday challenges. Our very construction of mathematics is the flip side of our construction of ourselves, where these dual constructions are both compatible with certain modes of practice preferred by the models of governance to which we are subject. Mathematical ideas then are not so much tangible entities to behold but rather specific manifestations of human experience that require a specific mode of human to experience them.

The consideration of wider theoretical resources that might be used within mathematics education research is in some ways prompted by the way in which the field of mathematics education relates awkwardly to its two constituent terms. Mathematics and education wave tenuously to each other from disparate conceptual

T. Brown, *A Contemporary Theory of Mathematics Education Research*,
https://doi.org/10.1007/978-3-030-55100-1_1

domains. Mathematics is considered by many as a discipline beyond social discourses. Underlying this understanding is a philosophical position that seems to assert the objectivity of mathematics as a prized possession. This kind of grounding has the effect of conceptualising mathematics as constituted by pre-existing patterns that are stable and can be discovered. In this view, it is possible to know what is true and what is not true since knowledge is objective and universal. In some versions of this formulation, the reality of mathematics has the same qualities regardless of observer and context. Philosophies of mathematics centred in positivistic notions of mathematical truth, objectivity and stable meaning are not especially disposed to the familiar philosophical bases of education. Nor do they resonate well with the more nuanced *linguistic turn* privileged in all three mainstream philosophical traditions of the later part of the century: hermeneutics, analytic philosophy and postmodernism. Truth, insofar as it is entertained in these philosophies, is processed through language where knowledge emerges through the operation of discursive systems often without centres. As such, knowledge houses tendencies that are not always in the business of portraying a world defined by consensual harmony in which final answers might be available. Research, in these contemporary traditions, is more centred on generating alternative analytical filters according to diverse priorities rather than supposing that a best solution could be achieved. Here, mathematics is fitted to specific purposes, where analytical structures are imposed *on* life to make life different, rather than being discovered *in* life that goes on much the same after the discovery.

Education is notionally a social science susceptible to interpretive analysis. Social science brings to the fore the complexity of the social world and that vantage point prompts the notion that people are constantly making sense of their worlds. Realities are local, specific and constructed, and, hence, values are an integral component in the meaning systems that people generate in social action. Truth, then, is not absolute and certain but is socially and experientially based, embedded in fluid social interactions. From this specific sensibility towards knowledge construction comes the understanding that the social world can only be investigated through a systematic analysis of socially meaningful action. Education, however, frequently resists conceptual immersion in the broader social sciences and the analytical resources those sciences provide since, as both idea and practice, it finds itself increasingly susceptible to external definitions and overt and covert regulation. Curriculum decision-making is split and shared unevenly between various groups that do not necessarily see eye to eye. Their differences result in disjunctions (both real and potential) between mathematics education policy setting, curriculum implementation by teachers and the conceptualisations of mathematics education by researchers. The resolution of these conflicts has resulted in an enforced homogenisation of pedagogical practices together with demands for increased testing. Research then needs to decide which side it is on, whether it is supportive of specific political agendas or resistant to those agendas, perhaps in pursuit of more intellectual ambitions, but then who would such intellectual ambitions serve? For example, is mathematics education research designed to support institutionalised conceptions of mathematics predicated on prescribed targets, or is it about taking a more critical approach to such prescription with view to opening new trajectories, such as seeing education in much broader terms, thereby challenging more the familiar framings that characterise the common sense of the day? Any answer to such a question is far from clear-cut. Institutionalised conceptions of mathematics are often created towards supporting agenda of inclusion, but that very inclusion may temper the aspirations of those wanting to pursue a more individual or eccentric path.

Restrictive conceptions of mathematics and of education, like these, mean that the composite term "mathematics education" is held in place by a variety of culturally bound assumptions. Largely circumscribed by something bigger than itself, mathematics education is constituted through dense webs of power. Traces of the determining effects of power are apparent in any mathematics education community of practice. A dilemma presents itself to those involved directly with those communities: Do we conceptualise our task in terms of initiating our students into existing knowledges? Or, might our task be seen, more radically, as troubling the limits of those knowledges, with a view towards keeping open the prospect of our students accessing a truth that transcends the parameters of our own teaching since the world that they are entering is one that we do not know ourselves? In other words, is it possible for students to reach beyond the frameworks that their teachers offer to produce a new future beyond our current vision? The latter option is not to be taken lightly since it requires a major shift in conventional thinking and practice. How do we fashion a new imaginary in which teachers forego a comprehensive understanding of what their students should be able to achieve? Thus, a key question for mathematics education research can be framed in this way: Is it possible to embrace new ontological possibilities for the learner and teacher beyond established states of representation? The intention here would be to open another space for talking about the field in a way that is responsive to the diverse demands it encounters and the multiple contexts that shape its practices.

This book seeks to provide a theoretical account of how processes of learning and teaching mathematics create us as particular types of human compatible with prevalent ideologies. Not so much inclusivity for all in the study of mathematics but rather compliance for all. The book speculates on why the mathematical work that precedes each of us motivates us to understand ourselves in the way that we do. But having understood ourselves in given ways, how do those self-conceptions then motivate us to construct mathematics in our own actions and pass it on to new generations? The book argues that caricatures, whether of humans or of mathematical ideas, result from contingent aggregations of historically derived elements. In these caricatures, we fix ourselves as "humans" by "counting as one" a certain set of elements (body parts, key locations, years of experience, grade point averages, Facebook "likes", consumer preferences, etc.). We fix mathematics in much the same way (multiplication tables, iteration processes, graphs) and cross-reference these reductions to each other to the potential exclusion of renewal seen in more nuanced terms. That is, compliance for mathematics and for people in the name of inclusivity according to current agenda with the chosen characteristics sutures new ways of being.

Mathematics as a field of human intellectual endeavour preceded all of us living today. We have learned to believe that mathematics can do a lot of things for us and we trust "it" with our lives. Many of us mundanely rely on it to keep its peace to hold up bridges and buildings for centuries. Just a select few people rely on cutting-edge mathematical innovations to enable sophisticated ventures like a short break to the moon. Ultimately, "we believe that it is linked to the fantasy of control over a calculable universe necessary to sustain our present social and political order" (Walkerdine 1988, cover text). So, mathematics has become an inextricable part of our lives, where strict boundaries between practical and intellectual manifestations of mathematics are difficult to draw. These boundaries are yet harder to discern since pedagogical interventions impose multiple understandings and levels of trust in "real-life" models. It is sometimes unclear whether in these attendant pedagogical rituals we are aiming to keep mathematics alive in its responsiveness to new challenges, or rather calcify old versions of life and the forms that it takes in the name of wider circulation today for pedagogical accountability.

Against this supposed backdrop of mathematics always having been there in ways that are familiar, we build an understanding of who we are. But how do we encapsulate who we are against this backdrop, using the paraphernalia of that backdrop as seemingly raw materials in constructing our story of who we are? Freud (2002, p. 5) argues that:

> An adult's sense of self cannot have been the same from the beginning. It must have undergone a process of development. Pathology acquaints us with a great many conditions in which the boundary between the ego and the external world becomes uncertain or the borderlines are actually wrongly drawn. There are cases in which parts of a person's body, indeed parts of his mental life – perceptions, thoughts, feelings – seem alien, divorced from the ego, and others in which he attributes to the external world what has clearly arisen in the ego and ought to be recognised by it. Hence, even the sense of self is subject to disturbances, and the limits of the self are not constant.

That is, my sense of self is always rather speculative. Lacan's iconic example of what he calls the Imaginary is that a child looks into a mirror and says, "That's me". But this identification is with an image, or caricature, rather than the real me. What's me, or not me? "The Imaginary is the transformation that takes place in the subject at the formative mirror phase, when it assumes a *discrete* image, which allows it to postulate a series of equivalences, samenesses, identities, between the objects of the surrounding world" (Bhabha 1994, p. 77). The Lacanian subject is known through the stories in which the subject appears, such as in a psychoanalytic encounter where an analysand depicts aspects of her life through a sequence of spoken words. That is, the focus is on how life is organised as a conglomerate of words or symbols or stories or narratives rather than on a supposition of an *actual* (biological) life to be observed and classified according to key characteristics. The signifier is privileged over signified. The story that is told somehow replaces the life that it sought to describe.

The notion of "one life", "one self" or "one individual", however, is not always quite so distinct. Research has described many examples of children accessing mathematics through computers where the boundary dividing teacher and student is obscured. For example, the teacher function in the educational use of software can be enacted in different ways with different degrees of human teacher input. It is easy to generate many alternative contemporary examples where the nodal boundaries (teacher, student, mathematics, human, machine) are rather less clear, such as between where the human stops and the machine begins: children sharing an app on an iPad; computers consummating a prearranged date to trade shares as predicted market conditions move into place; Andy Warhol getting confused between the real and the artificial; Lewis Hamilton and Felipe Massa who became renowned for repeatedly driving their cars into each other and blaming the cars; Arnold Schwarzenegger's alter ego terminating one of his adversaries; the absence of centrality in the World Wide Web; Stephen Hawking producing equations through his electronic media; or Richard Dawkins and his genes each claiming primacy. The talking and gesturing individual human described by Piaget as an immediately present physical entity is rather less prominent in the landscape of contemporary society with machines or pedagogical apparatus replacing so much of what had previously been more direct human contributions. These *machinic* supplements to human activity have earlier mathematical conceptions built into them, like bionic arms. The assumption of a self in an assertion of saying "that's me" comprises a collation of a set of characteristics, attributes, organs, etc. that make up "me", for now. This set of characteristics is "counted as one" person. Yet there are different ways of constituting "me", and different aspects of oneself create the characteristics that make "me". And in these constructions of myself I am using, knowingly or unknowingly, more or fewer of the machine-like supplements that are available to "me". My personal boundaries lack clear definition. And I can never be sure how much "me" integrates forces that I might not support in conscious awareness. Ian McEwan's fictional futuristic character Adam is a factory made entity who has so many human characteristics that he is unsure whether he "feels" human or not (McEwan 2019).

Technological advances have resulted in the very infrastructure we inhabit absorbing socialised mathematical framings from earlier era (Bastani 2019). For example, the widespread personal ownership of smartphones has relocated and redefined the very collectivism of encountering mathematics and the tangible manifestations or nodal points that locate and define mathematics in the popular imaginary. Numeric algorithms are absorbed into sequences of button presses, swipes, etc., whilst geometric objects are constructed and apprehended according to the processes of digital apparatus, rather than with analogic rulers and compasses. But these digital manipulations conceal design-stage choices in terms of how certain ideas or procedures are incorporated and understood. Pedagogical choices or functional routes have been made within the technology prior to the user pressing any buttons at all. Similarly, the very physical and mental formation of humans themselves is a function of the textualised and mathematical ecology of which they are part, and their choices feed into the big data that characterises new forms of normality. In some countries, smartphones provide an excellent means of governmental surveillance, where it can even be decided if someone deserves a holiday.

School mathematics is increasingly viewed as part of the apparatus deployed in responding to political demands for economic and technological development. Schooling in general, and mathematics education, is increasingly shaped, funded and judged by its perceived capacity to deliver success in terms of the prescribed quantitative measures by which so many governments reference their ambitions and achievements. Good performance here has sometimes been taken as being indicative of wider economic potential: the policy rhetoric suggests that the more we can improve in those areas, the better for our future national well-being. Governments of right and left have been seduced by the appeal of "raising standards" in a statistically defined world, in which standards become a fetish for intellectual life and academic achievement. Measures of school performance developed in various international exercises now often define what education is for or what it should be, policing educational boundaries with ever-greater efficiency. These instruments have transformed the content of what they purported to compare and similarly threaten to transform the demands on teachers and pupils preparing to meet these newly defined challenges. A key effect is a convergence of the metrics that produce normalcy, equating compliance with specific patterns of achievement with being "good" or "better", or even "outstanding". Policy thus legislates for a specific version of mathematics according to a centralised script, normalising what it is or should be to be a mathematics student and what it is or should be to be a mathematics teacher.

But "improvement" or "maximising" and similar aspirational metaphors for the passage of time can be understood in many ways. Academic motives and ethics for working with children in school such as enjoyment of mathematics, mathematical integrity and functionality in practical situations do not always pull in the same direction as "improvement" or its metrics. A choice needs to be made as to the sort of mathematical activity that is worth living, and what or who it is for or against. Do we want to invest funds in centres of excellence in learning at the expense of wider inclusion? Should mathematics be promoted at the risk of discriminating against certain students or promoting dominant political agenda? Should mathematical understanding be conflated with functional technology? We might even ask whether functional mathematics or its pedagogy is inhibited by overly asserted notions of certainty. Further, the *advance* of mathematics is not always desirable. Often the economic drivers of research in mathematics are not decided by altruistic purpose or ethical priorities. Missiles rely on research into sophisticated mathematical models and that can influence the priorities of government funding in mathematics. Our access to scientific and mathematical phenomena is mediated by multiple foregrounds and is affected by the way in which we apprehend their purpose and accept the challenge of engaging with them as imaginations, possibilities, obstructions, hopes, fears, stereotypes and preconceptions (Skovsmose 2016, 2019). Manchester residents Ernest Rutherford and Alan Turing each provided operational levers to ending World War II through their work in mathematics. Rutherford probably did not predict Hiroshima as an application of his work when he split the atom. Turing's work on breaking codes, however, is credited with shortening the war by 2 years, by weakening the Nazi naval siege of Britain. We might also add that mathematics is implicated in the ongoing financial uncertainties where confidence intervals have sometimes delivered their outliers. Bankers have calculated their bonuses, but not the outcomes of their own actions amidst the seismic sliding. Their sums seem not to work for other people. Ambitions to improve the teaching of mathematics can serve multiple ends, not all worthy of our support.

Barwell (2019) investigates how we might conceive of mathematics education in ways that is supportive of the environment, maybe in designing technologies that do not pollute. Ernest (2019a) reviews some of the wider issues relating to the ethics of mathematics.

Contemporary politics is complicated by the disjunction of governmental politics and the real operation of the market, which forces the hand of states to adopt certain forms of policy. We do not elect the people who are really in power. Thus, market conditions can often displace educational principles in setting the terms of educational practices. That is, it can be unclear how a researcher in mathematics education might seek to conceptualise the challenge of researching the field with a view to asserting some instrumental impact. Impacting on policy is not only unlikely, as politicians do not always listen to or connect with mathematics education researchers, but even if they were to be more attentive, the impact of any given policy is highly uncertain. However, this macro perspective evades many researchers in mathematics education who focus on their own local situations, without any specified ambition of scaling up for a wider population. The difficulty of scaling up has been the theme of a recent ESM special issue (2019).

A major challenge then is to rethink the breadth of mathematics education in resistance to reductive conceptions of mathematics and to critique mathematics education conceived of and (re)created in support of current models of economic production, technology and political administration, rather than, say, social welfare or epistemic motivations. The political climate has reframed how funded research in mathematics education is conceived, prescribed, evaluated and so conducted. Market metaphors abound in the language of improvement, with terms like progress, advance, quality, effectiveness, industry, competitiveness, performance and standards slipping easily off the tongue in much of the contemporary academic discourse. Hence, much research is often predicated on *improving* school achievement in standardised terms rather than merely *studying* it and understanding it. Proposals for funding typically must offer victory narratives, making promises of how research outcomes will provide specific understandings of education and so improve it. References to such discourses seem often to shape the activity of aspirational individual researchers. The superlatives used in the construction of these narratives, however, can sometimes disguise the differences between the multiply directed motivations of mathematics education researchers (e.g. for ethical practices, to understand more deeply, to disrupt or think differently) and the operational motives that guide their actions (e.g. securing funding, getting published, recalibrating practice, working towards a PhD, helping their students, etc.). The requirement that research should reach agreement with politicians and employers across nations might be a further stretch.

The proposed reorientation of research activity focus is a key task for theory, and theory development alone justifies its importance to the mathematics education research community looking for fresh ways to understand its activity. The field of mathematics education research is populated by people who are typically quite good at mathematics, usually located in higher levels of education. Their efforts are often predicated on raising standards in a competitive environment to ensure adequate capability across the population but possibly rather less on wider inclusion across the spectrum of educational needs and aspirations. There are relatively few mathematics specialists working at the primary level addressing needs at that stage of education. Mathematics at the primary level is often tackled by more generalist educators where the specificity and identity of mathematics education might be seen very differently. Issues of inclusion in mathematics often need to be considered at a structural level of putting appropriate curriculums in place rather than equipping individual teachers with pertinent skills. For this reason, this book is less concerned with operating in a functional way at any specific level of education such as teacher agency but rather more concerned with understanding mathematics from a more general educational perspective across the breadth of schooling where the administration of that socially oriented schooling process impacts on the nature of mathematics as we understand it and on how it is taught. The book asserts a new "social theory" where both of those words remain in transition where the book's purpose is to articulate the mechanisms of that transition.

1.1 Chapter Outline

Chapter 2 provides a theoretical discussion of how we understand mathematical knowledge. The theory presents rationality and belief as mutually formative dimensions of mathematics, where each term is more politically and socially embedded than sometimes depicted in the field of mathematics education research. The chapter considers alternative modes of apprehending mathematical objects derived as they are from this socially defined space. Two accounts of how a young child might learn to point at mathematical entities are presented, where alternative interpretations of this act of pointing are linked to conceptions of sharing understandings. This comparison then underpins a discussion of how mathematics is produced as entities to be acquired according to certain shared ideological schema that also shape who we are. The chapter's central argument is that rational mathematical thought necessarily rests on beliefs set within a play of ideological framings that partition people in terms of their proxy interface with mathematics. The challenge is then seen as being to loosen this administrative grip to allow individuals to release their own powers to generate diversity in their shared mathematical insights rather than being guided by conformity.

Chapter 3 considers some of the arbitrary curriculum or assessment criteria that operate in the social construction of mathematics in educational institutions. The advance of mathematics as an academic field is typically defined by the production of new ideas, or concepts, which adjust progressively to new shared ways of being. That is, mathematical concepts are created or invented to meet the diverse demands of everyday life, and this very diversity can unsettle more standardised accounts of what mathematics is supposed to be according to more official rhetoric. For example, the expansion of mathematics as a field often relies on research grants selected to support economic priorities. In schools, economic factors influence the topics chosen for a curriculum. In some countries, there is a shortage of specialist mathematics teachers that limit curriculum choices and restrict the choice of viable teaching materials, educational targets or models of practice advocated by research in mathematics education. Our evolving understandings of who we are and of what we do shape our use of mathematical concepts and thus our understandings of what they are. School mathematics has been reduced according to ideological schema to produce its conceptual apparatus, pedagogical forms and supposed practical applications. The resulting cartographic definition of mathematics steers the production and then selection of learners according to arbitrary curriculum or assessment criteria.

Chapter 4 provides a more explicitly Lacanian examination of how teachers resolve the pressures of working to curriculum demands. Centred in the doctoral studies of my Manchester colleague Peter Pawlik, the chapter considers how recent international developments in mathematics teaching have been influenced by what we see as the ideological notion of the mastery curriculum. Lacan's four fundamental discourses (master, university, hysteric and analytic) provide an analytical framework linking governance, institutionalised education and resistance. A case study of a teacher is used to illustrate how this discursive patterning is integrated into practice.

Chapter 5 describes some empirical research in both primary and secondary university teacher education. It considers how practices of teacher education impact on classroom practice by new teachers and thus shape the mathematics that takes place. The theme is explored through an extended discussion of how the conduct of mathematical teaching and learning is restricted by regulative educational policies that set the parameters of teacher education. Specifically, it considers the example of how mathematics is discursively produced by student teachers within an employment-based model of teacher education in England where there is a relatively low level of university input. It is argued that teacher reflections on mathematical learning and teaching within the course are patterned in line with formal curriculum framings, assessment requirements and the local demands of their placement school. Here, both teachers and students are subject to regulative discourses that shape their actions, and, consequentially, this regulation influences the forms of mathematical activity that can take place and be recognised as such, but where this process restricts the presentational options for the mathematics in question. It is shown how university sessions can alternatively provide a critical platform from which to interrogate these restrictions and renegotiate them.

Chapter 6 provides an account of my own mathematics teaching with student teachers and explains why I find teaching mathematics so exciting if it can be linked to the generation of multiple perspectives to be shared rather than the reproduction of a dominant view with prescribed pathways to this view. Some trainee teachers report on shared experience in a spatial awareness exercise concerned with exploring alternative apprehensions of geometric objects. Examples are provided of student teachers encapsulating their perceptions. The diversity of responses reveals alternative subjective positions each highlighting different qualities of the apprehended object. I have sought to show through my own teaching how mathematical challenges might be seen more in terms of students being supported in developing accounts of and gaining confidence in their own perspectives rather than meeting preset objectives.

Chapter 7 digs deeper into theory to consider further how the mathematical/human interface depends on the mutual dependency of how we understand mathematical objects and of how we understand human subjects. The apprehension of mathematical objects is examined through sessions with student teachers researching and critically analysing their own spatial awareness from a pedagogical point of view. The chapter is guided by the theoretical work of Alain Badiou whose philosophical model develops a Lacanian conception of human subjectivity and defines a new conception of objectivity. In this model, the conception of subjectivity comprises a refusal to allow humans to settle on certain self-images that have fuelled psychology and set the ways in which humans are seen to apprehend the mathematically defined world. The assertion of an object, meanwhile, is associated with finding a place for it in a supposed world, where the object may reconfigure that world in its very assertion. The composite model understands learning as shared participation in renewal where there is a mutual dependency between the growth of human subjects and of mathematical objects. Renewal is referenced to a diversity of ever-shifting discursive parameters that enable learning through negotiating the spaces within which we operate and the objects those spaces allow. Learning to teach then comprises developing sensitivity towards the discursive spaces that allow others to build objects. The chapter again provides examples from my own teacher education activities centred in addressing these concerns.

Chapter 8 documents aspects of the discussion that has taken place as a result of sociocultural theorists responding to my engagement with their work in my book *Mathematics Education and Subjectivity*. Specifically, the chapter responds to Wolff-Michael Roth's critical reading of the book. His reading contrasted my Lacanian approach with Roth's own conception of subjectivity as derived from the work of Vygotsky, in which Roth aims to "reunite" psychology and sociology. I argue, however, that my book focused on how discourses in mathematics education shape subjective action within a Lacanian model that circumnavigates both "psychology" and "sociology". From that platform, this chapter responds to Roth through problematising the idea of the individual as a subjective entity in relation to the two theoretical perspectives. In line with the broader remit of this present book, the chapter argues for a Lacanian conception of subjectivity for mathematics education comprising a response to a social demand borne of an ever-changing symbolic order that defines our constitution and our space for action. The chapter concludes by considering an attitude to the production of research objects in mathematics education research that resists the normalisation of assumptions as to how humans encounter mathematics.

Chapter 9 discusses at a more historical level how our conceptions of mathematics and of ourselves as researchers, teacher educators, teachers and students have been transformed through mathematical activity being viewed through the apparatus of schooling and international comparative filters. This model provides an example of how changing practices impact on the social construction of mathematics itself. The chapter argues that the fields of psychology and mathematics each describe realities that are consequential to past human endeavours or conceptualisations. Neither of these fields depicts stable truths.

Chapter 2
Reason to Believe in Mathematics

2.1 Introduction

The supposed wonder of mathematics is often lost in formal education that is accountable to the mechanical processes that govern our lives. Pedagogical mediations tend to shape mathematical ideas so that they can be more readily seen, tested or applied. Even university mathematics comprises alternative topographies anchored on selected objects or procedures prevalent in certain places at given points in time. Meanwhile, various styles of mathematical thinking have been created, selected or funded to support practical enterprises such as building roads, the effective analysis of economic models, everyday finance, etc. The relevance of these enterprises, however, ebb and flow as time goes by, and so do the forms of mathematics that are produced in support.

The realisation of this contingency troubles any account of mathematics existing *out there* in some absolute sense waiting to be discovered or being seen through the objects that have already been noticed. It frustrates any attempts to pinpoint the undeniable successes of mathematical thought through its visible effects in everyday practical activity. Yet it is commonly thought that there is something significant in mathematical thought itself, whether discovered or invented, that needs to be accounted for beyond its apparently stable everyday appearances in the physical world. Seemingly, such thought has properties and a precision that produce results unlike other symbolic frameworks. Our practical applications, however, cannot fix our mathematics thinking forever. Its empirical link to seemingly tangible objects ultimately slips away.

This chapter seeks to show how the supposed existence of mathematics beyond its appearances relies on a play of ideological perspectives and on the individual's understanding of the demands that they face. The cut between those included in and

---

This chapter draws on material first published as: Brown, T. (2016). Rationality and belief in learning mathematics. *Educational Studies in Mathematics. 92*(1), 75–90.

T. Brown, *A Contemporary Theory of Mathematics Education Research*,
https://doi.org/10.1007/978-3-030-55100-1_2

those excluded from mathematical activity has nothing to do with any supposed intrinsic qualities of mathematics but everything to do with how mathematical ideas are packaged for human consumption. The chapter begins by considering the production of mathematics as a cultural practice and the apparent social division of labour between those who do more advanced mathematics and those who have it done for them. It ponders on the impossibilities of precisely locating mathematical entities, caught as they are between alternative social constructions. The second section focuses on the apprehension of mathematical objects by examining how we might conceptualise both a person pointing at an object and the apprehended object itself amidst diverse contextual parameters. The chapter concludes by arguing that earlier mathematical conceptions are built into the human self-image clouding from view the earlier human construction of, or belief in, that mathematics. That is, the rationalities of mathematics are always being reconstructed anew by each generation according to their beliefs, ideological preferences or specific "societal mediation" (Roth 2012a, b) but where those attributes slip beneath one's own self-consciousness and disturb its capacity to be complete.

2.2 The Location of Mathematics

We often trust in a mathematics that appears to exceed us and perhaps everyone else. As a mathematically adept reviewer of an earlier paper put it: "I believe the four-colour theorem has been proved, even though I have neither the know-how, inclination, or time to verify the proof". Probably all people prefer to let others know or do mathematics for them to some degree. A division is created between those who know and those who feel they no longer need to as they defer to experts or machines. But even for more advanced mathematicians, there is great reliance on technology that conceals so many of the operations beneath its overt functioning.

Clearly, no single person could know the full extent of mathematical knowledge. Yet for many people in the wider population, this trust in other people, or in technology, to do their mathematical thinking for them is rather more fulsome. Only half of British adults it seems can achieve the level of an average 11-year-old and very often members of the other half expressed their emotional attachment to mathematics through pride in their deficiency (Paton 2012). This disinclination to participate has provoked widespread curriculum reform targeted at ensuring that enough of the populace achieves some sort of basic functionality in the subject. These strategies sometimes fall short of providing the beauties of abstract thought and can shape mathematical thinking according to a consumerist agenda in which the learning of mathematics is seen primarily as the acquisition of knowledge specified in predefined ways.

This polarisation of attitudes to mathematics is structural and endemic to the capitalist contextualisation of our practice (Pais 2013, 2014, 2015). It is a matter to be lived with rather than resolved through supposing that differences between people can be reduced. That is, the social structure shapes mathematical thinking according to current societal norms, defining human identifications with mathematics according to the divisions of labour endemic to the structure. For example, word problems are never innocent. They typically depict a Western rectilinear world described according specific practices, discursive genres, demographic makeup and economic priorities. This book is essentially arguing that it is not possible to see mathematics outside of its apparent manifestation in such structures but that formal education imposes further restrictions on top of these. The book also presupposes that it is also impossible for each of us to see ourselves outside of those structures.[1] Our reality (e.g. capitalism) always already includes us as part of it.

Is then mathematics generated in the mind? Or is it in a gesturing finger movement that activates a computer? Or is the mathematics *in* the computer? Does mathematics rely for its existence on the computer's output impacting on the interpretations of the community of mathematicians? Perhaps in the case of someone like Stephen Hawking, mathematics and physics live through their dissemination in best-selling populist texts (e.g. Hawking and Mlodinow 2010). In part, the existence of mathematics is underwritten by its materialisation in structures, processes and human action, as things that can be pointed at. For example, extensive work on gesture within mathematics education research has considered how mathematical entities are materialised in human activity. De Freitas and Sinclair (2012, 2014) explored the alternative productivities of gestural and diagrammatic evocations of mathematical ideas. They asked how the qualitative dimensions of mathematics were a function of such materialisations. That is, is mathematics the same when it is pointed to in a gesture as when it is encapsulated in a diagram? How do they differentially evoke mathematical objects and the (human) subjects creating them? We cannot readily draw a clear line between the human body and their operation of cultural machinery (Barad 2007). As a learner of mathematics, my sense of where mathematics is located is never finally resolved. Is it part of me or not? Have I made it? Have I pointed to it? Or, to use my country's colloquial educational parlance, has it been *delivered* to me as if it is a product brought to me by a supermarket van? These concerns prevent any final resolution of the issue of location. Mathematical ideas necessarily comprise a play of perspectives.

2.3 What's the Point?

How then might mathematics relate to the "real" world? To what extent can mathematics be referenced to ostensible objects? How is it possible for someone to point at those objects? Let's take an example arising from a discussion between some

[1] These *Symbolic* (rational) structures that shape our ideological space are within a knot of mutual dependency with our *Imaginary* (beliefs) of who we are, and the *Real* that defies all symbolisation. As Lacan (2000, p. 95) puts it: "The aggressive tension of this *either me or the other* is entirely integrated into every kind of imaginary functioning in man". I have explicitly discussed this play of perspectives in the context of mathematical learning (Brown, T. 2011).

delegates at the 2013 conference of *Mathematics Education and Contemporary Theory*. Julian Williams presented a critical analysis of a book by Wolff-Michael Roth and Luis Radford (Roth and Radford 2011). The specific concern related to how the teacher and student identify and share a mathematical object. Williams (2015) provided two alternative accounts of how a child learns to point at an object and by implication two alternative accounts of how an object is brought into being. He attributes the first to Roth and Radford. He argues that the second is more in line with Vygotsky.

Account One: An Infant Makes a Random Gesture

> an infant makes a random gesture that seems to the carer/parent as though the infant might be pointing, the carer/parent interpret it as pointing, and consequently reach out and give the infant the object pointed to, and thereby 'teach' the infant to point at desired objects. (Williams 2015)

Williams cites Roth et al. (2012, p. 69) on this point:

> A movement receives the sense of an action of a particular kind first by the culturally competent individual before this sense comes to be actualised by the child. In the example Vygotsky provides, there first is a random movement. The child does not know its cultural signification; it does not (yet) know to point. Rather the parent who sees the child move understands it as a pointing gesture.

Account Two: The Infant Has a Clear Purpose in Reaching Out

> the infant *initiates* the joint activity by having a clear purpose in reaching out to grasp some desired object, such as a dummy or shiny toy. It is this desire which is interpreted by a culturally competent carer/helper, one who empathises with the infant's frustration, and who is thus motivated to help the infant relieve their frustration by progressing the action towards fulfilling its goal. The infant has then to notice the carer's action, recognise its association with their grasping movement, and practice this on a number of occasions. The practised infant looks at a desired object, reaches out, 'points', and looks to the carer for their reaction. The evolution of 'reaching out to grasp' into 'pointing to indicate' as a means to grasp/act on an object-motive and achieve a desired outcome involves an internalisation of pointing as a communicative act, whereupon the child has 'learnt to point'. (Williams 2015)

In both accounts, the pointing relation involves an *algebraic* connection between a signifier and a signified. *This* hand position is associated with *that* object. I recall a seminar of the Association of Teachers of Mathematics led by Caleb Gattegno where he made this point in mentioning the example of a baby pointing to a fly moving on the ceiling. But which came first, the pointing hand or the object? In the case of a shiny toy, a dummy or a fly, the hand points at a pre-existing entity. An alternative however is for the pointing hand itself to bring the object into existence. The discussion gets more complicated if attention turns to how the teacher and student might point at an algebraic relationship *as* an algebraic relationship. There are many ways of understanding teachers and students and the subjective positions that could be assumed by them. There are also many ways of understanding algebraic relationships. That is, the symbolic objects of algebra can be framed (or cut) in different ways according to the subjective position being assumed in relation to them. In the teaching relationship that Roth and Radford are depicting the teacher is pointing at an algebraic relationship *as* a teacher, where the "object-motive" is for the student to *see* this algebraic relationship.

There are however some issues that prevent us from framing or evaluating this encounter in a definitive way. Williams raises the question as to whether the student is required to see the object in the same way as the teacher or if the object is transformed through the pedagogical encounter. Williams favours the latter and, further, where the object-motive is encapsulated within a particular curriculum structure that fixes pedagogical relationships according to some particular ideology of education. For instance, the current obsession with international comparisons shapes curriculums so that algebraic relationships are framed according to how they would be tested within that regime. That is, the ideological productivity of the pedagogical encounter affects the object in question by situating it within a context or frame that favours some interests rather than others.

A more radical alternative might question the wider framing that promotes and situates algebraic relationships as having a pivotal role. As Williams suggests, Roth and Radford's discussion is centred on the apparent expression of emotion by some very specific sorts of students in a laboratory classroom governed by a school structure that shapes the teachers, students and mathematics within it. There are other ways of doing it. Local administrations variously shape mathematical instruction to facilitate learning in particular settings where maybe children are taught differently according to their ability, customary choices prevail (class size, resource allocation, styles of teaching), mathematical teaching is differentiated according to social group, the resistance of adolescents to adult guidance results in teaching styles being shaped by the needs of classroom management, etc. More generally, teacher capabilities are dependent on a broad range of factors. For example, generalist teachers at primary level may be insecure or unknowledgeable about algebraic relationships. This could affect how students encounter algebraic relationships, perhaps through didactic approaches with reduced scope for exploration or through a published teaching scheme in which the teacher herself has a rather marginal pedagogical role. In short, the teacher function can be variously distributed where the bodily limits of the teacher do not coincide with the limits of this function. More widely, setting policy to bring about widespread adjustment to teacher practices towards raising "standards", or national test scores, is a persistent aspiration. Yet advisory groups, regulators, trainers, research funding agencies and potential employers work according to a variety of perspectives and priorities in terms of what a curriculum is intended to achieve. This variety of interest results in disjunctions between specifications of policy, implementation by teachers and the conceptualisations made of such implementations by researchers across many diverse studies. In short, the constitutions of teachers, students and mathematics are contingent on many factors. The act of pointing at an object can never be understood in a straightforward manner as both *pointer* and *pointed at* defy sustainable encapsulations. So, what are we pointing at when we are pointing at mathematical entities, successive layers of an onion or an inner kernel? It would seem that mathematical entities comprise the very play of perspectives on them.

2.4 The Production of Mathematics

Mathematics as a field evolves through reaching new generalisations in newly encountered conditions. Over a longer term, the absorption of mathematics into life results in the field of mathematics itself changing. Certain elements of mathematics have been touched more frequently by the need to support applications (e.g. statistical analysis of demographic trends). The field of mathematics itself has been marked out and prioritised according to how it supports practical agenda. Some aspects are much more popular than others for this reason and tend to be more likely used in everyday life or secure research grants etc. In a BBC radio feature, a professor of mathematics challenged a director of a government research grants agency by claiming that one could only get research grants for statistics at the time, such was the drive of supposed applications. Accordingly, mathematics itself had been reconstituted to meet evolving social priorities and criteria. The historical circumstances that originally generated mathematical objects are often lost. The objects may have become a part of who we are such that we are no longer able to see them.

Repetitive movements or geometrical constructions, for example, are routinely built into our physical landscape or buildings such that we do not notice them anymore as we become ever more accustomed to moving through them. Those ways of moving become part of who we are, e.g. navigating a townscape or shopping mall, descending stairs without tripping (Spanjaard et al. 2008) or moving through crowds in football matches. Similarly, we relate to mechanical equipment, craft or technology, as in driving a car, adjusting speed, direction, estimating journey times, noticing signs, using tram timetables or display mathematics/spatial sense in our computer use. Doing sports comprises the bodily experience of pacing a run or a swim, flipping for a tumble turn, timings, judging distance/position in darts or snooker, levers and force in hitting shuttlecock or ball at full stretch for an overhead clear, or judging how the throw up of the ball on serve impacts on the success and speed of the serve. Or appreciating the way football players pass the ball "into space" for others to run onto, noticing the "beauty" of a well-placed pass, etc. Meanwhile, similar spatial awareness develops for children playing in playgrounds, or with constructional toys, in counting games, etc. Similarly, we can think of mathematical constructs that have become common entities in everyday life. Circles, for example, have been featured in many of the stories that we have told about our world. We may feel that we have gotten to know circles from a lot of perspectives, which results in them acquiring a broad set of qualitative features. We use circle as a concept in building our world, and as a result circles become materialised or absorbed in the very fabric of our physical and conceptual world. Stellated octahedrons in contrast have been denied that level of intimacy and familiarity with humans (Fig. 2.1). Geometrically speaking there is no reason as to why one might be privileged over the other. Circles have been reified not because of any essential difference between them and, say, stellated octahedrons, but because of merely historical and political reasons. It is quite difficult to sort mathematical concepts according to which ones are empirically referenced like circles and those that are not so common in appearance or utility, such as stellated octahedrons or double helixes.

Mathematics exists as models of knowledge that sometimes support empirical enterprises, but ultimately, as empirical support, the models always reach their limits. We can never use words to precisely specify what mathematics is as such. Yet this realisation does not assist us much with understanding the predictive capabilities of mathematics, which have real psychic effects in more abstract mathematical analysis, and material effects in practical enterprises such as building bridges, the effective analysis of economic models, everyday finance, etc. There is something more significant to mathematical conceptualisation that *needs* to be accounted for. It has a precision and produces results unlike other languages. Mathematics can guide us or structure our thinking, but it does not fix our ways of making sense. Mathematics introduces polarities around which discourse can flow and which result in actual impacts on the physical and social world. Yet the possibility of mathematics as a complete system that supposes grounding in some empirical reference always slips away.

Fig. 2.1 Stellated octahedron

2.5 Rationality and Belief

We process reality by referencing our experiences to our preferred ways of telling our stories. Slavoj Žižek's favourite example is the cinema since it provides us with the story forms against which we can gauge the pleasures and disappointments of our own everyday lives. Žižek, a supporting character throughout this book, is a philosopher and major contemporary commentator rooted in the work of Lacan and Hegel. We could as easily see schooling as a similar arena in which we make sense of who we are. The rationality of school mathematics is contingent on how we prioritise and order the ostensible objects that we believe to be a part of it within the school setting. We recognise ourselves in the stories we tell, but not quite, and our attempts to get a better fit motivates our participation. Any apprehension of reality requires a subjectively located view to structure what we see. We need to ask where are we coming from in seeing things in the way that we do. And ultimately, *our assumptions* as to where we are coming from become part of reality.

The retroactive twists we make on our narratives of participation, and our attempts to stabilise them in some way for posterity, comprise the very production of reality. The lived experience of a sphere is condensed into a form of words, a set of symbols, some drawings or the articulation of a mixed set of emotions. The mimetic act of making sense through experience produces reality in the always already failed attempt to stabilise the world to scrutinise it. The *attempted* conflation of time produces a parallactic play of perspectives, a compression of points of view, which necessarily exceed my sense of self achieved in any singular perspective.

So then, are we pointing at mathematical objects in which we *believe* as a consequence of our empirical experience or do we know that they are there as a result of *rationally* thinking them into being consequential to making sense of that experience? The controversial British scientist Richard Dawkins (2006) has provoked much public debate through seeing scientific rationality and religious belief as being in opposing camps: "I am against religion because it teaches us to be satisfied with not understanding the world".[2] Yet reason and belief are not simply opposed to each other. A hermeneutic circularity is implied where beliefs produce rationality and vice versa. More radically, from a Hegelian perspective, "the object is always-already bound up in the complex mediating process of the subject's thinking it, and conversely, the subject's thinking the object is itself bound up in the object's very existence" (Davis, p. 14). "What we experience as reality is not the thing itself, it is always-already symbolised, constituted, structured by way of symbolic mechanisms" (Žižek 2011, p. 240).

Sverker Lundin brought my own production of reality to my attention after reading an earlier draft of this chapter. Sverker asked whether I, Tony Brown, as I understand myself, ultimately *believed* that there is a field of mathematics beyond all of

[2] Source Google. Dawkins was the University of Oxford's Professor for Public Understanding of Science from 1995 until 2008 (Wikipedia).

the ideological distortions. Such a belief appeared to be materialised in my mode of expression, perhaps through an over-casual use of words by force of habit, as if I had left some part of mathematics undisturbed by the ideological analysis, and then discussed distortions, desires, relationships, positions, etc. If this were to be the case, Sverker suggested (in conversation), mathematical objects would be analysed: "as *nothing more* than reifications of discursive practice – as the result of 'counting as one' a range of practices, the result of geographical invariance and chronological stability, the result of learning to relate to them as objects, etc". Yes, I am guilty as charged as I still am quite unable to exorcise past versions of self, which have made me the fully consistent academic who I am now ☺. A Lacanian account of the human subject has no aspiration to settle down with a final correct version: "Don't expect anything more subversive in my discourse than that I do not claim to have a solution" (Lacan 2007). Here we do not have a mathematical backdrop that gets distorted through subsequent usage. This unity never existed in the first place. It "is just a retroactive illusion" (Žižek 2014, pp. 49–50). Further: "nothing has been abstracted from any reality. On the contrary it's already inscribed in what functions as this reality" (p. 14). The reality of the pointing hand described in our first example is embraced by a symbolic universe, which is disturbed by its inclusion, whether it is a "random gesture" or it has a "clear purpose".

Mathematical thought derives from realities that are consequential to past human endeavours or conceptualisations. Many mathematical objects have an empirically motivated dimension; circles are motivated by naturally occurring phenomena, iterative processes model humans experience of progressively getting closer, statistics organises clusters of human information, etc. But often this empirical motivation underlying mathematical forms is lost in history, and we may not fully appreciate the implications of earlier empirical motivation for structuring our thoughts how they now influence our preferred or familiar ways of making sense. We also lose track of how past ideological/subjective perspectives are built into current ways of looking. For example, the concept of sphere may influence the way in which we mark out space. Indeed, is direct apprehension possible without historically derived markers (that bring with them their own past ideological priorities or contingent intuitive sense of how things work)? Or might we need to loosen our reliance on past structures (e.g. Newtonian physics) and experience space through alternative constructions (e.g. relativity in space, sub-aqua spatial dynamics, echolocation, quantum physics).

We reflect the symbolic universe, and it reflects us. Both Darwin and Einstein were great individual figures, but the novelist Ian McEwan (2012) in his Guardian article "The originality of the species" argued that they were each standing in evolving symbolic universes at particular moments in time that would deliver the results to someone or other in due course.[3] Ultimately, like Hawking or Dawkins, they have become iconic figures providing symbolic points of reference and

[3] Darwin was ultimately *fitter* than Alfred Wallace who simultaneously reached the same conclusions independently, and so it was Darwin who was *naturally selected* and *survived*! See: http://www.bbc.co.uk/news/uk-wales-21549079.

particular inflections that seek to stabilise ever-shifting discursive arrangements. They successfully "cut" reality into a particular time-dependent configuration that allowed a particular kind of subjective hold or brought a particular form of "master-ised" discourse into play (Lacan 2007, p. 103), that is, a style of discourse picked up by others that shapes the character of the field and privileges some points of view over others.

Mathematical entities are built *in* the human's own self-image as they reflect the human challenges for which they are created. Humans, however, are also a product of the worlds that they have produced, where a division of labour has arisen to reflect and serve the prevailing social administration. The mathematical entities that they have constructed are then built *into* the human self-image as they reference themselves to the world that they have created. A child may understand herself in terms of her shoe size, the number of dolls she has, her age or her maths scores. A teacher may understand himself in terms of his key performance indicators, tax rate, postcode or Prozac dosage. These parallactic self-producing and self-validating rationalities trap us into believing that there are universal realities (or rationalities) as to what it is to be mathematical and as to what it is to be human. What had been understandings of ourselves have become policing structures. Mathematical thought presented as a set of potential acquisitions has created the *belief* that there is something more tangible that assumes the quality of reality. Rationalities are then produced that are in line with those contingent arrangements or understandings of the world. The material points of reference that characterise school mathematics then support both a belief in mathematical entities referenced to contemporary societal structures and a contingent rationality that connects them. Rational mathematical thought necessarily rests on beliefs set within a play of ideological framings that sort people into types by limiting mathematical and pedagogical options. "Reason is, in a way, not more but *less* than understanding ... all we have to do to get from Understanding to Reason is to subtract from Understanding its constitutive illusion. ... Reason is Understanding itself in its productive aspect" (Žižek 2020, pp. 72–73, his emphasis). Understanding per se lacks the precision of the reason actually delivered by understanding. School mathematics, for example, built in a contemporary human self-image, presents not so much a distortion of "genuine" mathematical thought as a mode of thinking that serves to produce then select learners according to arbitrary curriculum and assessment criteria.

2.6 The Incomplete Production of Mathematical Reality Through Commodification

How might we contemplate the reality of mathematics? This is not a new problem and makes regular appearances in the mathematics education research literature. In a recent review of philosophies in mathematics (Ernest 2019b), Otte (2019, p. 61) expresses it thus:

> Mathematics causes some specific epistemological and educational difficulties. On the one hand, it must accept the merely representational character of all knowledge. On the other hand, mathematical knowledge claims a special truth status, compared to other knowledge.

Responding to Otte, Radford (2019, p. 7) concurs by adding:

> Without a doubt, Platonism has had a privileged seat at the table of the mathematicians. The mathematicians' ontological position that attributes to the ideal objects an existence independent of human labour certainly has consequences in the manner in which research is conducted. It is not the same to assume that you create something as to assume that you are discovering it.

Radford captures a common understanding of the history of mathematics where ideal objects have come into being without the assistance of humans. They have been discovered rather than invented. Meanwhile, Schürmann (2019, p. 241) argues that "the separation between mathematics and reality is an outcome of several shifts in historical mathematics discourse". As Alexandre Pais (in conversation, following Hegel) put it: "one feels as if one is discovering something that has always been there. That our mind is just a vehicle for the development of the mathematical idea".

Can we then be more precise in depicting how mathematical concepts intervene in more ideological constructions of reality, where forms of practice motivate specific understandings of mathematical concepts? In a famous debate, Richard Dawkins represented a rationalist camp that "raged against any kind of mystery in the cosmos, preferring instead to settle for a cold universe driven by the machine of pessimistic reason" (Žižek and Millbank 2009, p. 6). He was countering Alister McGrath, a professor of theology, who had posited a religious thinker governed by faith.[4] A second debate, however, between the theologian John Milbank and the philosopher Žižek led to an assertion that faith and reason are not simply opposed to each other (ibid). They each argued in different ways that the work of Hegel undermined any dichotomy between the mythical and the rational. For Hegel, "the object [or concept] is always-already bound up in the complex mediating process of the subject's thinking it, and conversely, the subject's thinking the object is itself bound up in the object's very existence" (op cit. p. 14). "What we experience as reality is not the thing itself, it is always-already symbolised, constituted, structured by way of symbolic mechanisms" (Žižek 2011, p. 240). Reality, in its very constitution, is ideological. Žižek (2011, p. 144) identifies three positions in Hegel's formulation:

> In the first, reality is simply perceived as existing out there, and the task of philosophy is to analyze its basic structure. In the second, the philosopher investigates the subjective conditions of the possibility of objective reality … [we ask where are we coming from in seeing it that way]. In the third, subjectivity is re-inscribed into reality … [our ideological assumptions as to where we are coming from become part of reality].

He provides the example of art: "Reality is not just 'out there', reflected or imitated by art, it is something constructed, something contingent, historically conditioned"

[4] Andrade-Molina et al. (2019) also discuss the relation between reason and faith.

(op cit. p. 254). In postmodern art, for example, "the transgressive excess loses its shock value and is fully integrated into the established art market" (op cit. p. 256). Similarly, mathematics describes realities that are consequential to past human endeavours or conceptualizations or commodifications.

Mathematics as a field is built in the human's own self-image through its expansion according to social agenda. As seen, mathematics can provide a structuring or formalisation of one's connections to the world. Commodified versions of mathematics have created the illusion that there is something more tangible in mathematical thought that assumes the quality of reality, supporting thoughts directed towards more settled arrangements of the world. These constructions become the currency used to measure and classify mathematical thinking. The need for accountability in mathematical learning results in specific transformations of the mathematical teaching and learning around commodified concepts.

How then do these concepts provoke our willingness to be governed by them? Althusser (1971) argued that schooling provided a key element of "state ideological apparatus", the mechanisms through which the state disseminated its models of preferred behaviours. According to Žižek (1989, p. 43), however, Althusser, "never succeeded in thinking out the link between *ideological state apparatus* and ideological interpellation":

> Althusser speaks only of the process of ideological interpellation through which the symbolic machine of ideology is 'internalised' into the ideological experience of Meaning and Truth. (Žižek 1989, pp. 43–44)

In our case, the link would be between the assessment structures that govern our practice and our belief in those structures. Whilst we may criticise the structures in theory, our practice is largely compliant. As indicated, mathematics in universities and in schools interpellate individuals but why? Althusser offers no explanation. Žižek contrasts Althusser with Lacan who posits some subjective space that exceeds ideological interpellation. In a Lacanian framework, the subjective experience of mathematics can exceed these ideological parameters because of individuals *practically* participating in the rituals of schooling. In subjecting oneself to the ritual of institutionalised mathematics, one is inadvertently materialising one's belief in it, and this belief creates a successful link between *ideological state apparatus* and interpellation. Meanwhile, mathematical thought will always exceed its specific commodified manifestations such as the concepts that are constructed for school and elsewhere. After Kripke, Žižek posits a notion of a "'rigid designator' - of a pure signifier that designates, and at the same time constitutes the identity of a given object beyond the variable cluster of its descriptive properties." (Žižek 1989, pp. 43–44). The name "mathematics" locates something that is more than the sum of its descriptions, thwarting any consistent account of what mathematics "is". Rather, mathematics is only accessed indirectly through descriptions of the activities taking place in its name. And the sum of those activities is not the whole.

In brief, mathematical productivity results from a play of ideological perspectives, where arbitrary (master) perspectives are selected to facilitate social administration but in so doing reduce mathematics by restricting the sorts of more personal insights that can be acknowledged in a school setting. The challenge that this chapter advocates to the reader is to loosen this administrative grip through his or her own preferred point of leverage (ballot box, adjusting teaching style, influencing curriculum decisions, political resistance, writing a paper for ESM, etc.) to allow individuals to release their own powers to generate diversity in their mathematical understandings rather than conformity.

His own position is suddenly made clear here.

Chapter 3
The Social Packaging of Mathematical Learning in Schools

This chapter is useful for our paper in justifying the connection between cultural arbitraries of curriculum and pedagogy and pupils' experiences of M.

Educational thought is undoubtedly ideological, but its application to mathematical ideas can seemingly anchor more radical ambitions. It is commonly thought that you are either right or wrong in mathematics with little space between. Educational research either informs improvement or it does not. As seen in the last chapter, the advance of mathematics as an academic field more generally, however, is defined by the production of new ideas, or concepts, which adjust progressively to new ways of being. That is, mathematical concepts are created to meet the diverse demands of everyday life, and this very diversity can unsettle more standardised accounts. For example, the expansion of mathematics as a field often relies on research grants selected to support economic priorities. In schools, economic or political factors influence the topics chosen for a curriculum. Our evolving understandings of who we are and of what we do shape our use of mathematical concepts and thus our understandings of what they are. Moreover, public images of mathematics pull in many directions that produce alternative conceptions of the field. These disparities of vision result in much variety in how mathematical concepts are materialised in everyday activity. They also point more fundamentally to the uncertain ontology of mathematics as a supposed field itself and its evolution according to the demands made on it. As discussed above, mathematics as a field is often thought to exist as a consequence of rationality or even as a matter of belief. Ideology, however, can shape notions of utility, rationality and belief. School mathematics, this chapter argues, has been reduced according to ideological schema to produce its conceptual apparatus, pedagogical forms and supposed practical applications. It has been transformed as a result of ever more pervasive formal demands in schools linked to the regulation of citizens as part of the ideological state apparatus.

Draws on material first published as: Brown, T. (2017) Concepts and commodities in mathematical learning. In E. de Freitas, N. Sinclair & A. Coles (eds.) *What is a mathematical concept?* Cambridge: Cambridge University Press. (pp. 189–202).

T. Brown, *A Contemporary Theory of Mathematics Education Research*,
https://doi.org/10.1007/978-3-030-55100-1_3

The chapter asks how we think about mathematics through the forced standardisation of ideological lenses and contemplates the forms that it takes. It suggests that mathematics as a field is built in the human's own self-image through its expansion according to social agenda. These self-producing and self-validating relationships trap us into thinking that there are universal realities as to what constitutes a mathematical concept. The chapter offers some insights into how Žižek's work extends Althusser's model of ideology as applied to revising our understanding of mathematics itself. It asks how we think about mathematics through ideological lenses and contemplates the forms that it takes in the hundreds of hours that it occupies in most people's school education. The subjective experience of those hundreds of hours experienced as part of the social management of populations may act in the service of those ideologies by making us believe them through the sheer force of habitual action. Here mathematics as a more reflective activity resides in a parallel universe available only to those able or prepared to temporarily sacrifice everyday life to pursue the beauties of more abstract thought and in so doing downplay analytical opportunities that could be more widely available through mathematical thought understood in a more inclusive way.

For various reasons, however, many people decline the benefits of a mathematical education. As mentioned, a recent report in Britain claimed that only 50% of adults function above the level of an average 11-year-old, and very often members of the other 50% were quite proud of their limitation (Garner 2012). The report's author, Chris Humphries, the chair of National Numeracy, was rather concerned: "Now that's a scary figure because it means they often can't understand their pay slip, they often can't calculate or give change, they have problems with timetables, they certainly can have problems with tax and even with interpreting graphs, charts and meters that are necessary for their jobs" (ibid). In addressing this problem, the wonder of mathematics takes on a very different style in many school contexts that are shaped by teachers' accountability to regimes designed to support the many and various mechanical processes or market structures that govern our lives. As seen, the pedagogical or practical mediation pertaining to such regimes reshapes and *commodifies* mathematical concepts into objects that can be more readily tested or applied within these regimes. That is, for many students in schools the space of mathematics is marked out by mechanical skills and procedures supportive of ideological agenda. These aspects are privileged over developing more intuitive powers or other aspects of mathematical learning. The pedagogical objects of school mathematics (multiplication tables, Pythagoras theorem, decomposition method of subtraction), however, still mark concepts that retain their structural place within mathematical thought that exceeds these ideological parameters, ways of mathematical thinking that are suggested beyond the bare symbols (e.g. conceptualising iteration to infinity, the sense of a rotation in an angle measure). These latter aspects of mathematics are "exempted from the effects of wear and tear [where the supposed field of mathematics itself] is always sustained by the guarantee of some symbolic authority" (Žižek 1989, p. 18). We forgive mathematics its awkwardness in everyday life as we sustain a faith in something more pristine.

So, although the very existence of mathematics is linked to our practical applications, there is also some implied claim to an underlying truth in a more abstract sense. Recent research in mathematics education has pointed to how the existence of mathematics is underwritten by its materialisation in structures and processes (Palmer 2011). It is never entirely clear where the human stops and where the operation of cultural machinery begins (Barad 2007). Was the mathematics that Stephen Hawking generated in his mind or in his computer? It is this sort of dilemma that has fuelled mathematics education research in recent decades. Research in the area had often in the past been governed by Piagetian conceptions of the mind (Piaget 1952). Children passed through successive developmental stages where it was the teacher's job to enable the children to reconstruct ideas as they followed the inevitable or "natural" route to maturity. Mathematics and the mind were two separate entities that got to know each other in the classroom. The international conference on the *Psychology of Mathematics Education* has provided a long-term centre of gravity for international researchers in the more generic field of mathematics education. In the last couple of decades, however, discursive constructions have become more familiar. In these later models, the focus is not so much on minds developing than on changing the story or structure that individuals follow. On the one hand, in this scenario students can construct their own accounts of mathematics, bringing new sorts of mathematics into being to meet the needs of their personal circumstances. In some contemporary understandings of mathematical learning, pupils investigate mathematics towards introducing their own individualised structuring of the landscape being encountered further blurring the line between the individual human and the mathematical concepts that she produces (e.g. Brown 2011). Conversely, on the other hand, policy makers can legislate official versions of mathematics, a more centralised script as it were, and police their implementation towards greater conformity (Brown and McNamara 2011).

3.1 Curriculum as Acquisition

School mathematics is understood through curriculum formulations, teaching schemes and textbooks that challenge children and teachers to follow or create mathematical procedures. These formulations are shaped around objects that could be pointed at and become more or less familiar objects through repeated use (e.g. fractions, Pythagoras formula, circles) or procedures (e.g. long multiplication, factorising). Certain styles of questioning are favoured, especially those that more readily lead to clearer assessment and specific modes of interpreting pedagogical encounters. These familiar mathematical forms have become the institutionalised markers of much school mathematics that promote conformity (Brown and McNamara 2011; Brown and Clarke 2013; Williams 2015). For example, multiplication tables often provide a key reference point in school mathematical learning and become part of the caricature of mathematics with which pupils identify. That is, pupils begin to understand themselves as being mathematical through indicators, such as their proficiency in learning their tables.

The literature in mathematics education research has widely reported on the caricatures or beliefs, which orient the pupils', or teachers', broader mathematical understanding, shape their experience of the "pedagogical encounters" that bring them together and, less often, of the notional "macro political context" that shapes their actions. Fischbein (1987, p. 206) argued "there is a world of stabilized expectations and beliefs which deeply influence the reception and the use of mathematical and scientific knowledge". Yet, Leder et al. (2002, p. 1) saw this dimension as a "hidden variable in mathematics education". Goldin et al. (2009) have reviewed the large volume of later work that has addressed this apparent deficit. Skott (2014, p. 3) has suggested that research could usefully "shift the focus from beliefs to the pre-reified processes that are said to give rise to them". For instance, our beliefs about school mathematics relate to rationalities, cartographies and codes of conduct produced through earlier beliefs. A recent international handbook surveys this theme (Potari and Chapman 2019). More broadly, the addition of elements to the school curriculum (e.g. tables and graphs) and the reduction of other areas (e.g. geometry) mark the ongoing historical formation of mathematical ideas in the context of social practices. Systems of rationality evolve with beliefs: "what others have learned has to be re-learned, re-integrated and re-expressed in each generation" (Mason 1994, p. 177); "the *being* of what we are *is* first of all an inheritance, whether we know it or like it or not" (Derrida 1994, p. 54). Ernest (2016) discusses this matter in detail through historically changing understandings of certainty in mathematics.

For example, school mathematical themes could be seen as being constituted through *counting as one* a certain set of objects (tables, graphs, etc.) (Brown 2011, b). But similarly, the points on the curve produced by the equation $y = x^2 + 3$ get counted as one and get to be called a "quadratic" so that the term "quadratic" becomes a particular enshrined object within the landscape in question that anchors or guides how we make sense (as discussed in a later chapter). Yet the statements that seek to locate and define mathematical phenomena so often become the statements that police its boundaries and set its policies on inclusion and selection. Pupils must know their tables and recognise a quadratic if they are to advance in their mathematical studies in school, as it is understood within the given regime.

Further, mathematics, as it is taught in many schools, is often referenced to mathematical content of the sort a university academic mathematician would recognise. This may provide the frame of reference against which the correctness of mathematics carried out by children in schools is judged. The mathematics encountered in schools is also locally defined around social practices, such as calculating supermarket bills, estimating the number of bricks needed for a wall, predicting climatic trends, etc. The point, however, is not to target the supposed underlying mathematics as the ultimate quest but rather to (1) question why mathematical activities in the classroom have assumed the social forms that they have and (2) to explore the consequences of those outcomes. The English mathematics curriculum has partitioned mathematical ideas and themes for consumption in schools. The British government has exercised its control over teachers and students by specifying specific skills and competencies, which stand in for the government's supposed obligation to promote a numerate population with consequent benefits to our society, technology and the economy. Maybe even those marks are not being hit and we could pursue alternatives.

Mathematical activity is commonly understood as being targeted at evoking specific mathematical concepts. Alternatively, however, the activity could be understood as a microcosm of social activity more generally. Mathematical activities governed by certain procedures, rules, performance criteria, etc. might be referenced to other social discourses, including others specifically related to mathematical heritage. As seen in this way, mathematical objects become a function of their relationship with multiple discourses. This softens any assumption that the activities are necessarily anchored in specific mathematical concepts. Rather, the reification of the supposed concepts is unfolded across multiple sites. Here there would be no universal conceptions of what mathematics *should* be about; rather our conceptions would be linked to the historical and social processes that generated classroom mathematics in the material forms it now takes. As suggested, the advance of mathematical thinking is defined by the production of objects, often in response to newly defined applications, funding priorities and pedagogical circumstances, or as a result of ever more pervasive formal assessment demands in schools.

Such trends are consequential to school mathematics being pulled in two directions at the same time. School mathematics serves mathematical heritage, and often mathematics education research understands its task in terms of serving that ambition. This orientation however obscures the demands of a more insistent master to be found in the political and economic structures that shape so many of our everyday actions, and in particular our encounters with what mathematics has become in Anglophone contexts at least. The pursuit of economic ambitions sometimes seems to result in school mathematics favouring the performance of skills and procedures rather than nurturing the student's more intuitive powers of mathematical rationality. Contemporary politics, however, is complicated by the disjunction of governmental politics and the operation of the market, which forces the hand of state to adopt certain forms of economic policy (Bauman 2014). That is, real power is no longer with governments setting policies, and the explicitly stated regulative apparatus that shapes school practice reveals that governments are merely acting in the service of oligarchic powers that transcend them, and over which they have little control in terms of delivering an equitable distribution of economic, cultural and educational capital (Piketty 2014, 2020). Pais (2015) takes the example of how motivation is activated between the two contrary demands and transcends much work on beliefs in mathematics education research by insisting on an overarching political dimension in linking mathematics to beliefs about what it is:

> (To) believe that mathematics as an object has already in itself the properties that will trigger students' desire for learning is to neglect all the students for whom engagement in mathematics does not derive from a "will to learn" but from a will to satisfy some Other's demand (say, parents' demand for good grades, teachers' demand for learning, academic or professional demands, etc.). It is an aspiration as pious as it is naive to assume that students will engage in mathematics for the satisfaction of exploring mathematics. To use the Lacanian lexicon, it is the cause and not the object of desire that determines students' engagement in mathematics. This cause has to be located not in intrinsic characteristics of mathematics nor in the innermost core of student's being, but in politics.

The mathematics that we encounter in schools has been shaped according to ideological schema to produce its pedagogical forms, schematic applications and the type of students it wants to include or exclude or can afford to fund or not (Lundin 2012; Pais 2014, 2019; Pais and Valero 2012; Pais and Costa 2018; Cabral and Baldino 2019). The assessment of school mathematics, for instance, is linked to the regulation of citizens as part of what Althusser sees as the wider *ideological state apparatus* "through which the symbolic machine of ideology is 'internalised'" (Žižek 1989, p. 43). Ideology speaks through us by processing what we say according to its preferred mode of shaping the world. It "is language that uses us" (Lacan 2007, p. 66). As Pais (2015) puts it: "ideology is not a distorted representation of a true reality. Ideology *is* the reality, we are 'naturally' in ideology; our natural, immediate, sight is ideological". But for Lacan the representations, or actuality, of school mathematics evoke rather than fully capture more embedded or historical mathematical understandings. That mathematics is supposed to be slightly beyond reach gives it a mystery and allure. For the advanced mathematician, the attraction may be the supposed abstraction that transcends the mere symbols. For the less enthusiastic student, it may be cool not to be a geek. The name "mathematics" is a "pure signifier that designates and at the same time constitutes its identity" and "locates something that is beyond the variable cluster of its descriptive properties" (Žižek 1989, p. 98) thwarting a consistent ideal account of what mathematics "is". Peeling away the layers leaves us with nothing. There "is no 'world' outside language, no world whose horizon of meaning is not determined by a symbolic order" (Žižek 2012, p. 366). As Lacan (2007, p. 124) puts it: "language … cannot be anything other than a demand, a demand that fails". Mathematics is *only* produced through activities taking place in its name, but this name has been linked to certain political preferences that do not reveal their true purpose.

3.2 The Production of School Mathematical Concepts

In the mundane of everyday life, children in schools learn or create mathematical procedures or pedagogical forms for handling different problems, such as those found in teaching schemes, textbooks or curriculums (e.g. how to teach the multiplication of fractions). The mathematics curriculum defines the forms through which school mathematical concepts are understood. For example, concepts of spatial awareness are learned through constructing triangles, reflecting shapes on graph paper, etc. Fashions change, learning theories move on and teaching schemes get replaced, resulting in school mathematics receiving regular makeovers whether or not these effect substantive changes (Brown 2012). School mathematics reifies certain objects (e.g. the first ten integers, the formula for factorising quadratics, BoDMAS,) or procedures (e.g. the decomposition method of subtraction, use of logarithmic tables) for greater scrutiny. It is often applied mathematics shaped around recognisable situations. Some configurations are repeatedly used resulting in the landscape of mathematics being viewed through perspectives that begin to characterise our engagement with mathematics. Questions are asked in familiar ways. Some areas of mathematics are favoured, identified as a specific skill, such as the concepts that are more easily tested (finding the difference between two integers, finding the area of a triangle) rather than exploring a mathematical terrain. Further, mathematical language, as used in schools, points to styles of social interpretation, social practices and ways of understanding the teacher-pupil relationship. As Pais (2015) suggests:

> mathematics education research is often looking at what Skovsmose calls a *prototypical classroom* and ignoring everything that somehow does not fit the picture of a well-organised and equipped class, with a teacher desiring to teach and students willing to learn. In much of the research into mathematics education, students and teachers are depicted as fully assuming the symbolic mandate conferred upon them. When problems appear, they tend to be ignored by research, or "solved" through the implementation of a better practice.

In Žižek's (1989, pp. 11–48) notion of commodity-fetish, commodities (or specific forms) become the supposed objects of desire. In our case in question, *commodified* versions of mathematics have become the institutionalised markers, or concepts, of school mathematics (Brown and McNamara 2011). The commodified objects, or mathematics' greatest hits, orientate our understanding of the subject. For example, if mathematics needs to include learning multiplication tables, the emphasis on mathematical tables becomes part of the commodification of mathematics and the way it is understood more broadly and how it is understood colloquially. That is, the table compiling multiplication results becomes an object, a "counting-as-one" of a certain class of results that provide points of reference orienting the pupils' wider engagement with mathematics – "do you know your times tables?". The addition of elements to the school curriculum (e.g. tables and graphs) and the reduction of other areas (e.g. geometry, proofs) marks the ongoing historical formation of mathematics in the context of social practices. But the statements or concepts that locate mathematical phenomena so often become the statements that police the boundaries of mathematics. Pupils must then know their tables if they are to advance in mathematics as it is understood within the given regime.

Whilst university mathematics provides a system against which the correctness of school mathematics is judged, the latter is more often locally defined around social practices, such as calculating supermarket bills, estimating the number of bricks needed for a wall, predicting inflation, etc. But why have classroom activities assumed the social forms that they have? That is, why have they become commodities with a given form? Commodification suggests a form of packaging designed for presentation in a chosen way of life where worlds have been conceptualised around them. This conceptualisation of alternative worlds built around commodities mediates the production of mathematical concepts and proliferates or typifies the senses in which they can be understood. The core mathematical idea may be linked to a way of life but in so doing normalises particular forms of life as though they were a transparent layer free of ideology. Language is using mathematics to process its preferred form of life.

For example, the English mathematics curriculum has "formatted" mathematics for consumption in schools (Skovsmose 1994). The government has exercised its control over teachers and students by specifying specific skills, conceptual awareness and competencies, which stand in for the government's supposed obligation to promote a numerate population with consequent benefits to our society, technology and the economy. Mathematics is characterised by the identification of a particular set of elements, which in turn imply a specific understanding of the world and how it might be changed. The route through which this can be achieved, however, may be difficult to specify in advance or interpret in retrospect. We may ask, what was in successive government ministers' minds in introducing such policy instruments into English schools? (cf. Žižek 2001, pp. 61–62):

- The minister wanted to improve mathematics by whatever means as part of her quest to provide an education as a basic human right – any rationalisation of how she achieves this is secondary to that basic desire.
- The minister saw pursuit of the improvement of mathematics as a good ploy for re-election – her only real concern.
- The minister sincerely believed that the implementation of her policies will bring about improvement in mathematics in the way she suggests.
- The minister was herself aware that policy setting is not an exact science but instinctively believes that a simple and insistent presentation of her policies will achieve for her the best possible outcomes in some way or other. This might be through good participation among teachers, quantifiable improvements in test scores, an image of a government taking charge or, more negatively, the demotion of mathematics as a political issue in the public's eye.

Which account best describes the minister's perspective? Perhaps all of them do. It seems impossible to attain a "real" version of events governed by straightforward causal relationships. And readings of what *actually happened* may change. The options above merely provide alternative fantasies through which reality might be structured. To personify the implementation of policies with a clear association between one person's rational action and its effect risks oversimplifying the broader concern. The effects of policy implementation are probably too complex to be encapsulated instrumentally, yet policy makers will nevertheless want to assert their instrumental powers or their supposed capacity to solve a problem, such as children failing in mathematics. As Pais (2016) argues:

> from the moment we introduce into school a promotional criteria, there will always be people who fail. ...That is, people fail not because the system sets them to fail, but because of some particularities that could be ameliorated through better research and school practices. This creates an entire academic and educational industry (from conferences, journals, or international assessment mechanisms to companies specialised in the production of didactical materials, teacher training or private coaching) aimed at solving the problem of failure in school mathematics. The focus is not the entire system where failure is necessary, but particular faults likely to be solved through expert engagement. The school's credit system is sustained by the illusion that success is possible, if only obstacles could be removed. What we fail to recognise is how these external hindrances are there precisely to create the illusion that without them, mathematics education will be possible.

From a teacher's point of view, one might contemplate reducing the emphasis on singular *metaphorical* associations between mathematical activities and mathematical concepts, in favour of a *metonymic* association between mathematical activity and social or political activity more generally. This entails linking the mathematical activities (seen as activities governed by certain procedures, rules, performance criteria, etc.) with other social discourses, including others specifically related to mathematics. The meaning of the mathematical discourses thus becomes a function of their relationship with the other discourses with which they are entwined, interpretive links that can always be revisited or renewed. This softens any assumption that the activities are anchored in specific mathematical concepts. Rather, we need to attend to the reification of such supposed concepts as they unfold in specific discursive environments. This would move us away from any supposed universal conceptions of what mathematics should be about; instead, it alerts us to the historical and social processes that generated classroom mathematics in the forms it now takes. School mathematics then presents not so much rational mathematical thought distorted by irrational beliefs but rather a specific mode of activity referenced to the performance of certain substitute skills and procedures that have come to represent mathematics in the school context consequential to the demands of social management.

From a student's perspective, mathematics can often be presented as though it comprises singular answers to any given question, as if there is always a right and a wrong answer. This view of mathematics promotes a pedagogical attitude governed by the commodification of objects characterised by *this procedure getting that result*, verifiable rather than true. Yet it is possible to produce mathematics as a conceptually defined space in different ways. As will be seen in a later chapter, in some of my own teaching, I designed some activities towards enabling the students to develop their spatial awareness as a prelude to a more formalised approach to geometry. What had been movements of the body became materialisations of one's comprehension of reality itself. The experience of the configurations became linked to how one felt at the time, a narrative of participation formalised for posterity, for the time being.

Chapter 4
The Ideology of Mastering the Curriculum (with Peter Pawlik)

Becoming a "mathematics teacher" comes with many expectations and demands within the discursive territory of the classroom where each teacher operates according to their own subjective preferences (e.g. Brown 2001; Brown and McNamara 2011). This chapter investigates everyday interactions between a student teacher and pupils. It places attention on the sense one student teacher makes of their immersion in social structures and asks how they experience and negotiate the various demands that are placed on them. Through a multitude of filters, individuals recognise themselves as teachers, responding to what they think is expected of them. For example, Radford (2018, p. 22) refers to a "unique individual who, through her engagement in social activities, continuously positions herself through other individuals in the cultural-historical world as an unrepeatable entity always in flux". Even though each individual is unique, they still play to the same rules, common cultural expectations and the social organisation of how teachers may be expected to conduct themselves. The reality of teaching, however, is often a far cry from the fantasy of a rational and idealised teacher described in policy documentation. A training year may entail a steep learning curve in which student teachers transfer their attention from mathematical pursuits such as learning about finding the gradient of a curve to confronting real 13-year-olds with an anxiety towards adding fractions. This chapter draws on the experiences of Emily, a student teacher, providing insights into ways in which the idea of a mastery curriculum can shape ideological understanding of mathematics education in school and university settings in England. It explores the discursive construction of the *mastery curriculum* using Lacan's notion of the master signifier, as exemplified in declarative assertions of "how things are". This analytical tool provides an approach to disrupting habitual thinking patterns within regulative scenarios and opening alternative discursive avenues.

This chapter draws on material from:
Pawlik, P. (2020). *The discursive construction of the mastery curriculum in mathematics*. Doctor of Education thesis. Manchester Metropolitan University.
Brown, T., Rowley, H. & Smith, K. (2014). Rethinking research in teacher education. *British Journal of Educational Studies*. *62* (3), 281–296.

T. Brown, *A Contemporary Theory of Mathematics Education Research*,
https://doi.org/10.1007/978-3-030-55100-1_4

4.1 Mastery Teaching

As in many countries, recent educational reform in England has been informed by evidence from high-performing jurisdictions, in particular Shanghai. As mentioned earlier, schooling around the world is becoming increasingly shaped and judged by its perceived capacity to deliver success in terms of international competitiveness linked to economic agenda, often as indicated through performance in comparative tests. It is no surprise then that, in 2016, the Department for Education in England issued a press release, "South Asian method of teaching maths to be rolled out in schools" (DfE 2016). Subsequently a range of interconnected policy initiatives that promoted East Asian practices was formulated under the banner *teaching for mastery* (TfM) (Boylan et al. 2018). Although the word *mastery* does not appear in the 2014 revised National Curriculum documents for practice in English schools, the influential government-funded National Centre for Excellence in the Teaching of Mathematics (NCETM) has adopted the word "mastery" in relation to professional development and the teaching and learning of mathematics (NCETM 2016, 2018, 2019a). Masked by the appearance of independent actors, government-funded agencies, such as the NCETM, promote specific classroom practices with an emphasis on East Asian styled pedagogical approaches (Boylan and Adams 2019). It is evident that the *teaching for mastery* strategy advocates particular teaching structures and methods. For example, a key component is the use of variation theory that includes multiple representations of what a concept is and what it is not (e.g. Kullberg et al. 2017; NCETM 2019b; Watson 2017). However, on Twitter in 2015, Mike Ollerton, an influential member of the Association of Teachers of Mathematics, argued that many of the features of *mastery* teaching have been around for 40 plus years. Irrespective of policy commitment, various official and unofficial discourses conceptualise *mastery* learning. That is, the current *mastery* rhetoric in England is a product of social cultural mediations, a conglomeration of approaches that is packaged by contemporary educational policies. Many schools are adopting a *mastery curriculum* influenced by East Asian approaches. They often appear as little more than reconceptualisations of earlier models of educational practice. The archaeology of the term *mastery*, for example, can be traced to the work of Benjamin Bloom, such as, requiring that pupils achieve a level of mastery in prerequisite knowledge before moving forward to learn subsequent information (Bloom 1968). Skemp's (1976) work on relational and instrumental understanding retains currency in the ongoing debate on the *mastery curriculum*. Singaporean approaches draw on Bruner's forms of representation, concrete-pictorial-abstract (Bruner 1966). Nevertheless, the NCETM, which operates with and through the Maths Hub network (Boylan et al. 2018; Maths Hub 2020), provide a framework and act as vehicles or tools to implement particular pedagogical approaches and government policy. Alongside the NCETM the government founded Education Endowment Foundation (EEF 2017) produced a report on improving mathematics for children in the age range 9–13. The report endorses many of the components of teaching for *mastery*. Even though there is a drive to centralise the content of professional development through the work of national organisations such as the NCETM and the Maths Hubs,[1] there are other providers and business systems which have set up versions of the *mastery curriculum*, for example, White Rose Maths and Mathematics Mastery[2] (e.g. Mathematics Mastery 2020; White Rose Maths 2020). As such, the *mastery curriculum* is packaged as a product, with curriculum materials, professional development, and a range of expertise and so on. This product or idea of *mastery* provides a framework for mathematics teaching and learning, which functions as an ideology that can provide a point of reference or identification for teachers giving a sense of collective purpose, a toolbox of pedagogical strategies and administrative procedures. That is, the vocabulary and language of the *mastery curriculum* can provide the orientation through which one recognises themselves as mathematics teachers. In a similar way, the teaching standards function as an ideology that can provide a point of reference for being an "outstanding" or "good" teacher and so on. Learning or teaching effectively in terms of the *mastery curriculum* only demonstrates subscription to that ideology. It only determines successful mathematics teaching if one buys into that ideology.

4.2 Teaching for Mastery: The Master of Us All

Whilst teachers are depicted as having agency within the neoliberal market-driven forces, teacher autonomy is reduced in the process of *statification*[3] (Boylan et al. 2019). That is, on the one hand, discourses of a free market and competition may seem to offer notions of choice and the autonomous individual. On the other hand, discourses of accountability and improvement do not leave much space for teachers to explore and be more expressive (Fielding and Moss 2011). As Clarke (2012, p. 48) argues this, "can be read as the subordination of this same self to the 'other', who determines and dispenses knowledge in the form of mandated curriculum, and who monitors its achievement through test and targets". As such, *mastery*, sold as good practice and as a means to improve standards, has an impact on the conceptualisations of mathematics education. It can be seen as a means of directing and controlling the actions of teachers and learners. Particular values and ideals of the *mastery curriculum* are presented as an absolute truth or, in Lacan's terms, as a master discourse (e.g. the supposed need to reduce the attainment gap and raise attainment). The emancipatory narrative being difficult to resist shapes the future of school mathematics.

[1] Maths Hub (2020). About Maths Hub. [Online] [accessed on 5 June 2020] https://www.mathshubs.org.uk/about-maths-hubs/.

[2] Mathematics Mastery (2020). *Secondary Classroom Resources*. [Online] [accessed on 22 May 2020] https://www.mathematicsmastery.org/classroom-resources-secondary-maths-teaching-support?phase=secondary&c=5ec7deb239305.

[3] Statification resonates with Foucault's (2009) "governmentality" in which the state appears as a given entity which is necessary for governmental practices to function.

Government rhetoric is difficult to refuse; the NCETM occupies a position of agency, and it represents systematic knowledge that addresses schools and teachers to enact and reproduce the knowledge system. The NCETM, among other agencies, could be described as disguising an authoritarian discourse with rationality. This systematic knowledge, the *mastery curriculum*, operates on the subject's desires to fit in, be successful, subordination to the perceived demands placed on teachers. Independent providers such as Mathematics Mastery might appear to offer spaces for professional development and teacher autonomy, but they too could be described as disguising the dominance of government discourse.

4.3 Ideology of the Mastery Curriculum[4]

The Marxist philosopher Louis Althusser (1971) regards schooling as one of the institutional ideological state apparatuses (ISA). For example, *mastery* teaching as a pedagogical tool is seen as a means to drive up standards. In this way, it looks like, "this is the way teaching has to be". As suggested in Chap. 3, in subjecting oneself to the ritual of institutionalised mathematics, one is inadvertently materialising one's belief in it, and this belief creates a successful link between ideological state apparatus and interpellation. Interpellation here can be understood as the subject feeling valued, fitting in within the establishment of the imaginary domain. Fundamentally, individuals are called into being through prescribed registers and discourses. The *mastery curriculum* places specific demands on individual teachers, to teach in a particular way. As such, the *mastery curriculum* is resourced with a kitbag of ideological state apparatuses (professional development, information, resources and so on). In establishing the self in relation to such discourses, the student teacher is interpellated as a particular subject. However, Brown et al. (2006, p. 33) argue that this sort of interpellation can be "delusional through its failure to embrace the whole picture". These ideas reverberate with Lacan's view that fantasies are deluded, characteristic of the imaginary.

Althusser maintains that cultural forms of ideology are constructed on an "imaginary relation to their real conditions of existence" (Althusser 2014, p. 181). However, Althusser differs from Lacan in his discussion of subjectivity. The idea that interpellation brings the subject into being suggests that the interpellated subject does not assume a prior conscious standpoint. As such, Althusser downplays the fragmented nature of repressed desire and thus displaces the role unconscious forces have on everyday actions. "Every subject endowed with consciousness/a conscience and believing in the ideas that it inspires in *her or freely accepts should 'act in accordance with her ideas' and therefore inscribe her own ideas as free subject in the acts of her material practice*" (Althusser 2014, p. 185). The assumption here is of the teacher as a rational agent, self-conscious and able to make clear judgement

[4] We thank Snezana Lawrence for framing this notion.

of ideological practices. For example, some teachers might subscribe to the government rhetoric that the *mastery curriculum* will foster a radical shift in mathematics teaching and improve England's performance in the PISA rankings. However, at the same time, other teachers might be sceptical about the motives of that ambition. Yet both groups comply with the master discourses and find that their practice is defined by the *mastery curriculum*. The successful implementation of *mastery* policy is not necessarily an improvement in standards but by convincing teachers, teacher educators and the public that this version of teaching and learning mathematics is the correct one. Standards from this point of view have not changed, but the parameters through which successful mathematical teaching and learning is understood have. Althusser's theory of ideological state apparatuses is unable to account for individual agency and the complex interplay between the fragmented, desiring subject. Whereas Lacanian theory of the subject and its notions of desire underscored by the imaginary, the symbolic and the real orders (see Brown 2008c or 2011, pp. 119–125) unpick how a student teacher responds to the ideology of the *mastery* policy. That is, "a subject desires an object not due to its particular characteristics but because of the place such an object occupies within their libidinal economy" (Pais 2015, p. 380). In other words, the desire for teaching the *mastery curriculum* is not in its applications but the *desire of the Other* (Lacan 2006), the symbolic network that signifies the master curriculum.

4.4 Lacan's Schemata of the Four Discourses

The theoretical ambitions of this chapter are aimed at building a sense of how alternative discursive priorities variously work through teacher educator practice. The chapter takes the premise that motives are harnessed by identification with particular discourses (retention of university values, the need to support practice, the promotion of research, the need to comply with directives to retain "outstanding" status, etc.). Analysis of the data to be presented in the next section will examine how these identifications link to particular modes of practice (e.g. the assertion of the academic dimensions of training, the development or retention of humanistically defined pedagogical processes, the smooth operation of administrative frameworks, etc.). Meanwhile, policy documents define the parameters of teacher practice to the extent that participation in teaching and teacher education becomes a form of bureaucratic compliance monitored by an inspection regime that insists upon this taking place. Such identifications and compliances, however, may result in some emotional cost to the individual with associated awkwardness. Yet there is some chance that the individual may succeed in regaining some personal composure through formulating a more systematically considered response to these conflicting demands.

Lacanian psychoanalytic theory portrays a subject divided between what she is doing and what she says she is doing. This division is located differently for different people, and the type of division determines who you are, who we are and how power and displeasure/pleasure function to secure alignment or nonalignment with particular discursive formulations. The individual is constituted according to the composition and mode of their identifications. Lacan's conception of society is dominated by the practice or use of language, where "when I say use of language I do not mean we use it – it is language that uses us" (2007, p. 66). Further, "discourse can clearly subsist without words. It subsists in certain fundamental relations which would literally, not be able to be maintained without language" (2007, p. 13). He continues: "nothing has been abstracted from any reality. On the contrary it's already inscribed in what functions as this reality" (2007, p. 14). Žižek (1989) contrasts Lacan's notion of a divided human subject with Foucault's late work, which was concerned with articulating the different modes by which individuals assume their subject positions. In Foucault's analysis, the subject creatively surfs from one subject position within a discourse to another to produce different effects, to craft a technology of self. Whereas Žižek (1989, p. 175) suggests that Lacan focuses on a subject who exceeds discourses, "the failure of its representation is its positive condition"; that is, the human subject thrives through not being pinned down in a clear definitive statement, leaving personal space to resist regulative impositions.

Lacan's schemata of the four discourses are referenced to systems of knowledge (university); discourses of control or governance (master); the alienated or divided subject split between alternative discursive modes (hysteric); and systematic resistance to oppressive power structures (analytic). This Lacanian model has been discussed in detail in the context of secondary school English education in a book by Brown et al. (2019). For this chapter, the schemata is drawn on in conceptualising how mathematics teachers craft their sense of being with reference to the discursive orders that determine their subjectivities. It provides a helpful model in depicting the "schizophrenic" subject positions that teachers are obliged to confront. For example, the individual will form identifications with political, academic or administrative discourses which shape that individual's thought and affect enjoyment and the meanings that he or she assigns to different situations. For example, the individual will form identifications with political, academic or administrative discourses which shape that individual's thought and affect the enjoyment and the meanings that he or she assigns to different situations. It is through this route that the chapter will theorise how the changing policy environment variously impacts on individuals and how they understand their mode of participation.

4.4.1 *University Discourse*

The university discourse comprises systematic knowledge. For individuals to understand this discourse, they need to be receptive to the idea of pre-constituted knowledge. This requires that the individual empties "themselves of any knowledge that might interfere with the knowledge in the discourse becoming an amorphous, non-articulated substance … to be articulated by discourse" (Bracher 1994,

p. 109). They are produced as a divided subject as a result of this interpellation that captures part of them; for example, a teacher educator is appreciated merely to the degree that their practice complies with inspectorial criteria. It "is admissible only insofar as you already participate in a certain structured discourse" (Lacan 2007, p. 37), but part of their selves is left out in this encounter, a gap, marking the divide. In turn, others may gauge the degree of this individual's submission according to particular criteria and judge their performance according to their degree of alignment. For instance, a trainee mathematics teacher may be assessed in their ability to teach fractions in a step-by-step fashion according to a curriculum schema that specifies particular developmental stages of a child's learning. Other aspects of their teaching, such as their humanist mode of interaction, may not register on this scale. A new entrant to the profession of teacher education, meanwhile, might be able to play one version of university discourse off against another (e.g. conceptual versus procedural knowledge) as teacher education boundaries lose definition. Emily, in her reflections, specifically challenged the system of knowledge when teaching directed numbers using the context of a witch's cauldron: "moved onto the cauldron analogy, with hot coals and ice cubes. I found this easy to explain, however it was difficult to get the students to use this idea as they had a very fixed knowledge of the 'rules' in their mind, e.g. 'two minuses make a plus'".

This production of the divided subject, however, is not the whole story, as Lacan portrays systems of knowledge as being in the service of alternative master discourses shaping the situation in question: "the master's discourse can be said to be congruent with, or equivalent to, what comes and functions … in the university discourse" (2007, p. 102). That is, the subjective production results from participation in a form of knowledge that is motivated by some underlying interest (mode of sponsorship, pedagogical preference, kinship, etc.).

4.4.2 Master Discourse

Alternatively, we could centre our attention on master discourses directly. Neoliberal trends have resulted in governments around the world shaping education according to economic conditions (Zeichner 2010). The British government might be seen as operating particular master discourses in the service of its policy ambitions to reshape education according to market parameters. This discourse works through demanding compliance to certain operational or administrative protocols in the name of customary or desired practices. In Lacan's framework, which draws on the Hegelian master-slave dialectic, master discourses are selectively linked to particular elements of wider (mythical) knowledge. The "master's knowledge is entirely autonomous with respect to mythical knowledge" (Lacan 2007, p. 90). The master merely asserts a particular version of reality, as though it is supported by systematic knowledge, "master-ized" discourse as opposed to "mastered" (2007, p. 103). "It is all about finding the position that makes it possible for knowledge to become the master's knowledge" (2007, p. 22). For example, UK Schools Minister Nick Gibb (2016) described changes in mathematical learning as "We are seeing a renaissance in maths teaching in this country, with good ideas from around the world helping to enliven our classrooms". It is the idea of the *mastery curriculum* that is being sold, not the *mastery curriculum* itself. His assertion was seen in some quarters as producing a mismatch with reality where things are not quite how we are being told to see them, releasing space for questioning or resistance. Moreover, behind this notional master is a split subject suppressing aspects of reality in the name of asserting a clear instruction. Politicians sometimes place great importance on being seen as "very clear" to avoid any charge of weakness or confusion, perhaps through fear that it might undermine their capacity for governance. They are obliged to suspend doubt and make decisions to select one form of systematic knowledge rather than another, which by "virtue of its very structure, masked the division of the subject" (Lacan 2007, p. 103). In doing this, however, "he does not know what he wants" (2007, p. 32) or what he will get in return. There is a gap between demand and response.

4.4.3 Hysteric Discourse

Meanwhile, the individual may successfully act according to the master discourse. Yet there is a similar gap between performance and the awareness or articulation of that performance. Žižek (2006a, 2006b) argues that ideology operates through the maintenance of this gap between alternative identificatory modes. For example, Brown (2008d) depicts a head teacher exploring her own complicity in policy rollout as she moved between resisting policy intellectually and implementing it faithfully in a material sense. Similarly, "relationship maintenance" (Ellis et al. 2013, p. 270) might be viewed as an insidious way of getting university tutors to act in line with the required behaviour on route to teacher education having the lower university input to be discussed in the next chapter. The tutors may protest vocally but nevertheless materialise their own oppression through their very actions in supporting schools. Their actions in turn equip schools with the wherewithal to replace universities whilst disenabling university tutors from protecting their patch through their more traditionally defined skill base. It may, however, be that the individual begins to sense this gap. The hysterics discourse might be seen as being provoked in the subject by a confusing element intrinsic to the demand being expressed in the master discourse. The respondent may be troubled by the demand, a niggling feeling perhaps. What do you want of me? I must protest as this does not seem right! "Why am I what you … are saying that I am" (Žižek 1989, p. 113). The subject addresses the master, and the mismatch between demand and response hints at an aspect of knowledge that the master discourse has concealed. The subject had been spurred on by the niggle marking, a gap that had provoked unease with being completely compliant with the demand being made.

4.4.4 *Analytic Discourse*

Lacan's (2007, p. 70) notion of the analytic discourse is modelled on a Freudian psychoanalytic encounter, where the "subject of discourse does not know himself as the subject holding the discourse". Analysis is directed at disrupting or resisting master discourses enacted in the service of oppressive regimes: "this master's discourse has only one counterpoint, the analytic discourse" (2007, p. 87). One goes into analysis with the intention of discovering the unconscious forces that interfere with conscious actions or the gap between them. For example, alternative systems of knowledge may conflict with each other and cause disturbance to the subject. The analyst addresses the subject with a view to identifying the master discourses working through them. Through this process a master discourse can be revolutionalised, turned over, as the analytic resolution works itself through. "Knowledge then, is placed in the center, in the dock, by psychoanalytic experience" (2007, p. 30). The analyst address is underpinned by systematic knowledge, which is ultimately referenced to new coordinates – that is, held in place by new highlighted features that Lacan calls master signifiers (2007, p. 92), for example, an earlier UK education minister declaring that "standards must improve".

4.5 Emily's Negotiation of the Mastery Curriculum

Emily is a student following the 1-year Postgraduate Certificate in Education for secondary school mathematics, and her story begins in her second school placement. As a pre-cursory to her placement, she spent 2 weeks at university looking at alternative pedagogical strategies. For example, students attend lectures titled, "making sense of algebra". These lectures attend to using contexts that are designed to motivate, engage and develop conceptual understanding of pupils (e.g. Hough 2012). Subsequently when Emily starts her second main teaching placement in a secondary school, she is "impressed" with the school's approach to teaching mathematics (which is quite a contrast to her didactical experiences on her first placement). The school has recently decided to incorporate a *mastery* approach, and as such, a considerable amount of time is spent within the department talking about different pedagogical strategies. Whilst many of the students are challenging, Emily's initial reflection at this school is full of enthusiasm:

> A mastery style of teaching mathematics is promoted in the department, with a priority at KS3 (11–13 year olds). After spending some time on this whilst at university, I am truly impressed by the teaching style. Whilst this is more in depth and requires more time, my ideas and teaching have changed significantly. I find myself picking up on very small elements of language and proof that I would not have noticed before. I can see the benefits of teaching students the 'why' and 'how' some abstract concepts of mathematics is useful and can be applied. I have stripped back my own knowledge of maths to then reteach in another way.

Emily's reflective writing is presented as a discourse in which she is forming herself as a particular type of teacher. She is constructing her identity (or being interpellated) in response to the ideology of the *mastery curriculum*. Emily has an image of the teacher she wants to be. Spending time both at university and with her mathematics department has organised her desires and might explain her motivations to become a particular type of teacher. That is, Emily is making a link between sessions at university and her school's approach to mathematics teaching. The discourse of *mastery* teaching is strengthened as it is endorsed by Emily's identification with the idea of an ideal teacher. "So, this is what is expected of me", a teacher that asks "why" and "how". In her writing, she comes across as a unified subject, one that has no resources for resisting ideology. However, turning back to Lacan, we can develop a more complex account of both ideology and the subject of discourse.

In Lacan's framework, a particular agency is "only a temporary subject effect resulting from a temporary subject position, and in addition, subject structure is not stable" (Alcorn 1994, p. 30). In the discourse of the university, placing "succeeding in mathematics" in the position of truth allows us to understand the possibility of the *mastery* teaching as the agency of the discourse. This systematic knowledge, this way of "understanding mathematics teaching", addresses the subject. In doing so, *mastery* teaching offers an idealised vision of the complete teacher. Emily strives to teach in this way; she is even attempting to "strip" back her own knowledge of mathematics so that she can teach this way. Here the discourse of the university is having a "totalising and tyrannical effect" (Bracher 1994, p. 115). In doing so, systematic knowledge functions to enact or reproduce *mastery* teaching.

Emily states "my ideas and teaching have changed significantly". Meaning is produced by language, which is "driven or operated by subject-functions such as desire, temporality, repression, the imaginary" (Alcorn 1994, p. 24). Lacan (2007) proposes that discourse functions such as ideology or knowledge that operates upon the subject. Desire might be expressed in relation to the type of teacher Emily strives to be, "teaching students the 'why' and 'how'" of mathematics. That is, the Lacanian subject is connected to the realms of the imaginary, the real and the symbolic. These unique subject functions "produce the subject's particularity of discourse- a singular style of discourse that characterises the subject" (Alcorn 1994, p. 37).

As Emily learns new knowledge, she is motivated to change and modify her actions and even her identity as a teacher. Smith (1988) notes that the Lacanian subject can never be equivalent to a particular composition of knowledge but is operated by many layers of internal organisation. All these layers form a system, but the many parts of system are never fully configured, and this means that the subject can never purely be one thing but be the divided subject. In the case of Emily, the discourse component of the "mastery style of teaching" is the agency operating on her, but it is not a simple reflection of the discourse system but through a synthesis of remembered discourse (in part a history of discourse). This echoes Lacan's theory of subjectivity where "the subject operates discourse" (Alcorn 1994, p. 28). Emily is not a mere reflection of the discourse, but through her subjectivity, she manipulates and transforms the discourse. In this way, there is the possibility in the production of original discourse, new knowledge.

As the teaching placement progresses, Emily is more concerned in building relationships (in particular with her year 10[5] class); issues of classroom management are a constant concern. In the Lacanian framework, a rotation of the schemata offers possibilities of new understandings. The next extract highlights some of her concerns and anxieties:

> I feel more in control of the students-particularly the targeted students in year 10 that caused me issues. I spent a lunchtime detention with them talking things through and getting to know them. This has definitely helped and improved my relationship with them. I hope that moving forward this will continue and I especially look forward to parents evening next week meeting their parents.

At this moment in her teaching, the emphasis is on building relationships with her students. Emily's teaching of mathematics is taking second place to relationship maintenance (Ellis et al. 2013). She is positioning herself and her students within a particular power discourse. Emily is finding that student desires take priority over master demands, even if they are anti-productive. These are producing real tensions, and the power relations manifest themselves in Emily giving a lunchtime detention and "looking forward to parents evening". At the same time, however, she acknowledges that she needs to talk and listen to the students. She is forming herself as a particular kind of teacher in which herself and students acquire specific identities.

If we place Emily the divided subject in the position of the agent (discourse of the hysteric), this disrupts the authority of the master discourse. The "hysterical structure is in force whenever a discourse is dominated by the speaker's symptom" (Bracher 1994, p. 122). That is, her concerns and anxiety about the behaviour of her pupils manifest as a failure of the subject. There is a gap between what she thinks is expected from her as a teacher and her awareness of the performance. The wish for security and stability is helping Emily develop identities for herself and students. In the search for meaning and security, the subject responds by providing a new master signifier, in the form of a secure meaning that will overcome anxiety and give a sense of control, stability and respectable identity. It is thus, as represented in the schema, the production of new master signifiers, covertly producing a system of knowledge. In other words, the hysterical discourse challenges Emily's position and identity as a teacher; in asserting control and imaginings of improved relationships with students, she is producing a new system of knowledge; this is what it means to be a teacher; this is what the society expects from me.

The discourse of *mastery* teaching demands compliance to certain operational protocols in the appearance of making progress. In this way, the concept of *mastery* teaching resonates with the Hegelian master-slave dialectic. In short, "the master's knowledge is produced as knowledge that is entirely autonomous with respect to mythical knowledge" (Lacan 2007, p. 90). The discourse of the *mastery curriculum* presents a version of mathematics as though this is the only way to teach mathematics. As Williams (2019, p. 2) argues "the policy [Mastery mathematics] becomes the master of us all, and we are obliged to suspend our critical faculties and comply".

[5] Year 10 includes students aged 14–15 years.

We can analyse Emily's remarks by considering subjectivity, knowledge and modes of resistance. For example, to what extent is Emily resisting the knowledge of the mastery curriculum through her need to control the students? On the other hand, she is also motivated by knowledge and is able to resist previous beliefs by acknowledging the need to "(strip) back my own knowledge of maths to then reteach in another way". This suggests that as Emily learns new "knowledge", she is motivated to change her actions, values and even identity. However, Alcorn maintains that a Freudian perspective would suggest that, "subjects can never use knowledge in a disinterested way because knowledge is always intertwined with the structure of subjectivity" (ibid, p. 35). That is, the subject always produces knowledge. The intimate link between knowledge and human interest was the central theme of a classic text by the major social theorist Jürgen Habermas (1972). His *Knowledge and Human Interests* emphasised how "'subjective interests' do not stand outside social totality, they are themselves moments of social totality, formed by active (or passive) participants in social processes" (Žižek 2020, p. 104). Assimilation of knowledge is essentially connected to subject structure. In essence, it would mean a Capitalist is always predestined to be Capitalist, a Marxist forever condemned to be Marxist. How can a being be anything other than what one is positioned as being? This would be a disappointing extrapolation from the theory. Some forms of knowledge seem more independent of subjectivity than others do, for example, performing algebraic manipulation, multiplication tables and so on. These forms of knowledge seem less problematic to transfer and less prone to subjectivity. However, other forms of knowledge have a stronger relationship to subjectivity, for example, ethical, political and so on. Alcorn (1994, p. 36) argues that "while it is difficult, it is not impossible to achieve knowledge in these fields". That is, through political resistance (discourse of the analyst) but also being attentive to the features of analytic resistance, it is possible to develop knowledge implicated in the structures of subjectivity. In this way, student teachers would acquire more agency in their practice.

The analytic discourse disrupts the demands of the master signifiers by thinking about it repeatedly and hence lessening its intensity by gaining insight about its workings. In this case, resistance to the master is motivated by knowledge and self-consciousness. Change in the subject is possible as the ego "'processes' discourse and 'learns' to respond differently to the insistence of the unconscious" (Alcorn 1994, p. 40). That is, through the recognition that reality is an illusion, the subject gains insight of who they are not, with new possibilities for the self.

A reflective and critical stance towards teaching encourages student teachers to analyse their practice and hence lessen the intensity of master signifiers imposing centres on discursive exchanges. Taking a critical stance provides an interrogatory position, within which the subject can (as far as possible) unsettle the dominance of initial identifications and is open to new possibilities. As discussed earlier, Emily in reproducing the master signifier (mastery approach) reflects on the behaviour of her pupils, in effect disrupting the dominance of the master discourse; through this process the master discourse is reworked, and ultimately a new version of the master signifier is developed.

Emily's early experiences of a "mastery style of teaching" were progressive, to the point where she felt the need to "strip back my own knowledge of maths to then reteach in another way". In this way, Emily is being interpellated in response to the ideology of the *mastery curriculum*. As the discourse of the *mastery curriculum* unfolds, there are fleeting moments of clarity, "this is the way things are". However, these moments are temporary; the discourse is already moving on. There is often a gap between attempting to meet the ideology of *mastery* teaching and the reality of classroom interactions. Student teachers and in general the subject desire to close the gap between the fantasy of *mastery* teaching and the reality in the classroom. Other demands and factors influence what we do. As Walshaw points out (2008, p. 124), "closing the identity gap is what learning to teach is all about".

In repeatedly mapping out classroom interactions to different permutations of discourse, we generate different possible understandings. That is, "it allows us to understand the functioning of different discourses in a unique way" (Fink 1995, p. 129). In particular, it provides insights into the formulations between knowledge, master signifier, divided subject and otherness. It combines in one model, psychic structures, motivation, with semiotics and discourse. In considering the various positions of the master signifier, we produce different understanding of how the subject engages with discourse.

Chapter 5
The Social Administration of Mathematics Subject Knowledge Through Teacher Education

5.1 Introduction

Understandings of mathematical subject knowledge for teachers inevitably respond to changing environmental conditions. Specifically, school mathematics is a function of the administrative constraints prevailing in the given educational context. Such situations will be characterised by preferred styles of presentation, specified targets, etc. As we have seen in the previous chapter, teachers are typically obliged to follow curriculum guidance within such constraints in deciding how to teach or otherwise meet the customary practices in their place of work. Meanwhile, their understandings of themselves are a function of the demands that they perceive being made on them. What is expected of them in their given professional role? Student teachers and new teachers are especially susceptible to the guidance of others. They may not, however, be fully aware of how their actions are shaped by their identifications with the discursive landscape. How then might we make sense of the mathematics that takes place in the classroom when it is enacted, perhaps unreliably, through the teachers' mediation of external demands? This chapter addresses this question by considering some of the ways in which mathematics is discursively produced by Student teachers working towards meeting the demands of externally produced definitions of practice. A key assumption of the chapter is that mathematics as understood in mathematics education research cannot be understood separately to the way in which it is processed by teachers and students in the given situation. Mathematics is a function of its location and the way in which people are working mathematically in that location.

This chapter draws on material from the following publications:
Smith, K., Hodson, E. and Brown, T. (2013). The discursive production of classroom mathematics. *Mathematics Education Research Journal. 25*: 379–397. By kind permission.
Brown, T. (2017). The political shaping of teacher education in STEM areas. In J. Clandinin and J. Husu. *Sage international handbook of teacher education.* New York: Sage.

T. Brown, *A Contemporary Theory of Mathematics Education Research*,
https://doi.org/10.1007/978-3-030-55100-1_5

The chapter is referenced to a sequence of recent empirical studies of teacher education carried out by the author, reported in two books that were not specifically about mathematics education (Brown 2018; Brown et al. 2019). Some models of school-based teacher education in England were considered in terms of how they generated understandings of teaching and learning mathematics and some other subjects. Prescriptive policies prevalent in that country have resulted in ever-changing pressures on teachers to meet centralised criteria targeted on developing the practical skill needed to implement a detailed curriculum. This has led to some very specific interpretations of mathematics and its teaching. Teacher professional identity has been referenced to skill development within this frame and the wider assessment culture. The teacher's capacity to exercise professional autonomy has been shaped by these constraints. This scenario was discussed in detail in the context of primary mathematics education in an earlier book by Brown and McNamara (2011).

For nearly two decades now, student teachers in England have typically spent much of their training period in schools. An early government-initiated "employment-based" model of teacher education begun a decade ago and had student teachers located primarily in schools "learning from our best teachers" (DfE 2010, p. 23). In this development, student teachers who were more mature or highly qualified worked in a paid professional capacity from the outset of their "training". These newer models coexisted with the then mainstream established models where more time (but not much more time) was spent in university in line with government requirements for time spent in school. This chapter specifically discusses how student teachers participated in that employment-based model but references this discussion to wider conceptions of teacher education now prevalent within the country and beyond. The two recent books depict a more contemporary situation. The purpose of this discussion here for a wider mathematics education audience is to consider how conceptions of learning and teaching mathematics change through training being located primarily in schools. It is not being suggested that readers try this model at home. That is, the purpose of the chapter is to explore in this instance how mathematics is a function of the discursive environment in which it is encountered. The chapter investigates how student teachers identify with specific discursive framings of mathematics teaching pertaining to this model of training. It asks how school mathematics is understood, *empirically*, by student teachers following this route into teaching. These issues are contemplated through the eyes of university teacher educators who were obliged to conceptualise their professional contribution from within a rather marginalised role. From this perspective, the chapter provides a window on how teacher educators and student teachers variously conceptualised school mathematics and how these conceptualisations were influenced by multiple prescriptions, interventions and environmental constraints. It analyses the resultant conceptions of mathematics revealed by student teachers in their understanding of the challenges they faced.

This attention to a specific example, however, is directed at opening a more general discussion. That is, the chapter addresses the much wider question of how school subjects in any situation are a function of discursive parameters and how the language being used formats activity in those subject areas. In the case to be discussed here, the way in which mathematics is administered in the specific pedagogical environment determines what mathematics is. Having been determined in this way, those conceptions of mathematics can police the practices that have been developed in the name of mathematics. Nevertheless, we shall consider how teachers can develop the capacity to engage critically with this discursive environment in their place of work and beyond through building reflective research within their practice.

5.2 The Discursive Shaping of Research in Mathematics Education

Earlier work on the theme of mathematics education and language often addressed how mathematical language is spoken or written in everyday life or more particularly within a classroom environment (e.g. Brown 1997, 2001; Morgan 1998; Pimm 1987). Later studies have taken a range of perspectives on how language filters or produces mathematical understanding. Barton (2008) showed how mathematical meanings are a function of the specific language or culture. Another New Zealand study looked at how computer media impact on the hermeneutic processing of mathematical ideas (Calder 2012; Calder and Brown 2010). Brown and Clarke (2013) conducted an international survey of how mathematical understanding is shaped by its institutional context. Much research has focused on how discursive formulations shape conceptions of classroom practice and of the people working within them. For example, professional teacher identities are a function of how teachers understand themselves fitting in (Black et al. 2009; Klein 2012; Walshaw 2010). Conversely, Nolan (2016) asked how prospective secondary mathematics teachers were subject to official pedagogical discourses embedded in classrooms. Walshaw and Brown (2012) conceive subjectivity in terms of participation. Walls (2009) and Llewellyn (2018) each focused on children's subjectivities. Discursive elements also underpin conceptions of identity centred on "legitimate peripheral participation" in "communities of practice", derived from the work of Lave and Wenger (1991). For example, Solomon (1998) examines mathematics as a community of practice and the teacher's role as epistemological authority in inducting pupils into such practices. Goos (2005) provided a sociocultural analysis of the development of pre-service and beginning teachers' pedagogical identities as users of technology. Jaworski (2019) incorporates the notion community of practice into her discussion of inquiry-based practice in university mathematics teaching development. Watson and Winbourne (2008) edited a collection of work on this theme. Brown and McNamara (2011) considered student teachers as *subjects* in accounts of their own practices and how policy discourses were articulated through these accounts. The authors sought to understand how mathematics, primary pupils and teachers were shaped by policy initiatives and how they were included in the world depicted by the policy apparatus. For example, the government, rather than mathematicians or teachers, determined the constitution of mathematics within a legislated curriculum. Pedagogical discourses have been shown to govern the choice of teaching devices, which in turn condition mathematical learning. For example, mathematical texts conceal conceptions of the pupils and teachers for whom they are created. Dowling (1998) showed how tasks designed for "less able" students in a teaching scheme were different to those given to "more able" peers. For any given topic, the emphasis in instruction varied between the texts, resulting in exclusion for the "less able" from the real business of more abstract mathematical learning. Instead, they were caught in the discourse of "less able" mathematics characterised with associated styles of illustration, questioning and assumed perspectives. The activity materialised the children as "less able" as they were doing the things deemed suitable for "less able" children. Meanwhile, Cooper and Dunne (1999) showed how "realistically" contextualised test items designed for greater accessibility (and with a certain sort of pupil in mind) in fact produce greater class and gender differentiation. Working class children were less able to spot the "game" of school. Wagner (2012) considered how students are constructed in school texts but also how the texts replicate teacher positioning and voice.

5.3 International Changes in Teacher Education

Many recent policy initiatives in teacher education have been consequential to the recasting of mathematics as a subject conceived as an aid to economic development, rather than, say, social welfare (Atkinson 2018) or epistemic emotions (Muis et al. 2015). A review of research in mathematics education covering the last two decades identified two prominent lines of research, one more theoretical concerned with identifying and codifying practices of teaching in general and the other more specifically practice-based pedagogies (Charalambous and Delaney 2019). There have been at least two very different state-led responses to changing teacher preparation designed to "improve" achievement. In some countries teacher education increasingly comprises a vocational employment-based model of training located primarily in schools. England is a prominent example (Brown 2018), with similar models being introduced in New Zealand and the United States. This approach is in sharp contrast to models followed in continental Europe, where student teachers spend much more time in university. "Almost all [European] countries introduced reforms in initial primary teacher education after the initiation of the Bologna Process (1999)" (ENTEP: Dimitropoulos, online), similarly for secondary subject teachers and half of pre-primary sectors of education. The model was motivated by sharing good practices and creating mutual trust in the teaching qualifications awarded across member states with a view to enabling shared accreditation and greater mobility across European countries. For example, in Spain, all primary teachers study at university for 4 years, including short periods in school, or 5 years in Finland where a master's degree is required for secondary teachers. The lengthy academic training often conducted by people with relatively little experience of schools, however, can seem distant from the more practical challenges ahead. In Germany, for example, teachers need to get through 4–5 years prior to being admitted to the school practicum phase of 18 months to 2 years. Yet this investment of time in university retains wide support across European nations. As one German primary mathematics teacher educator put it, "The university is a space to question. What for? Why? How could it be different? Rather than being in a state of permanent emergency (as in school-based work) … A teacher is not just a craftsman". This intensification of the academic component is a further distancing from practical concerns for student teachers in those countries (Hudson and Zgaga 2008). Once qualified, however, following an extended school placement after the academic component has been completed, rather more professional autonomy can be asserted by classroom teachers in Germany in making local decisions and setting the curriculums than in the policy-dominant approach in England.

These two approaches, school-based and university-based, reveal radically different conceptions of how teacher quality might be improved in the name of international competitiveness. In the first, teacher education has been wrested from its traditional home within the academy where universities play a support role to what has become "school-led" training where government funds for teacher education have been diverted to schools. Teacher professional identity has been referenced to skill development within this frame and the wider assessment culture. The second model, meanwhile, is similarly concerned with "raising teacher quality … (but specifically) in a way which responds to the challenges of lifelong learning in a knowledge based society" (Dimitropoulos, Ibid). It is characterised by reinvigorated faith in academic study and promotion of individual teachers, where a pedagogical dimension is included from the outset of undergraduate studies, but with relatively brief periods spent in school.

5.4 Changes to Mathematics Teacher Education in England

University mathematics teacher education in England has been redefined through new priorities determined by, among other things, budgetary constraint, problems with teacher supply (Rowland and Ruthven 2011; Williams 2008) and perceived school performance as compared with other countries (DfE 2010). The teacher education function has been redistributed to include professional and subject mentors within the school setting (cf. Jones and Straker 2006). These mentors are themselves classroom teachers with their own classes to teach. This arrangement is thought to provide immediate opportunity for student teachers to develop classroom skills (DfE 2010). The student teachers spend much less time at university with tutors, where they have some limited scope to reflect on their practice and to consider educational theory. Some research, for example, has focused on the importance of teacher reflection in university settings and providing the resources for teachers to creatively generate mathematics in productive classroom exchanges (e.g. Brown and Coles 2012). Space for such activity has been greatly reduced.

Hitherto, little research has been carried out on how increased school-based training supports the mathematical aspects of teacher education and how they are conceptualised, prioritised and enacted, so that further interventions could be better informed. We know little about how new teachers understand mathematics following training across school and university settings and how student teachers conceptualise their own teaching of mathematics in schools.

My own study provides an up-to-date overview of the current state of affairs (Brown 2018). Teacher education provision has become largely dictated by market conditions with some institutions better able to retain control over the content of their courses. But within many courses, something of the order of 30 days[1] is spent at a university during a 1-year postgraduate "training" course, where the chief university responsibility is oversight and accreditation for a process primarily administered by schools. University teacher educators and school mentors, however, may have very different priorities for their roles in teacher training, such as those relating to how subject knowledge is understood, meeting the demands of testing, effectively using materials, learning a range of pedagogical strategies or building personal involvement in the subject. There are different ways of understanding the disciplinary knowledge that teachers need. Schools may prioritise the immediacy of classroom practice or following centralised guidance; some (but not all) universities may prioritise the more intellectually based elements such as pedagogical subject knowledge, building professional autonomy or meeting the demands of formal qualification (Hobson et al. 2009; Hodson et al. 2010). Hitherto, relatively little research has been carried out on how increased school-based training supports the pedagogical subject knowledge aspects of teacher education and how they are conceptualised, prioritised and enacted, so that further interventions could be better informed. Meanwhile, the tendency in some countries to take charge of school practices through a multitude of regulatory devices, such as through frequent testing, prescriptive curriculum and school inspection (Askew et al. 2010; Brown 2011), has resulted in mathematics subject knowledge becoming understood through a culture of performativity (Pampaka et al. 2012). This insistence on following centralised "masterised" documents (doing what they *should* be doing) has deflected attention from knowing how the redistribution of teacher education has resulted in student teachers *actually* understanding and meeting the professional challenges they face. These changing policies affect the challenges faced by teacher educators and "school mentors" and in turn influence student teachers' conceptions of subject knowledge and its teaching. The policies also impact on the identity of the student teachers. Are they student teachers engaged in an educative process developing the ability to lead curriculum initiatives as they later become professional teachers? Or

[1]As a mathematics educator, one might like to estimate how many hours within these 30 days in university are spent on specifically mathematical themes for (a) primary trainees and (b) secondary maths trainees. The answer will be revealed on the next page. The government insists on a certain number of days in school during the year leaving relatively little space for universities to expand their provision. Meanwhile, it also insists on some of the time in university being spent on more generic themes such as safeguarding and British values.

are the trainees fulfilling the requirements of training, working to the current models of school practice, as specified by the government? That is, are teachers curriculum makers or curriculum implementers (Clandinin and Connelly 1992; Schwab 1983)? One is moved to suggest that school-based trainees are being prepared primarily for the latter and will take their chances in being carried along in the future rather than having been prepared directly to address changes in professional circumstances.

Such lower-cost school-based teacher education may appeal to an increasing number of governments in building and influencing the practice of their teaching forces. But three questions immediately present themselves: Does it provide a viable alternative to university-based teacher education? Does it alter the composition of the pedagogical subject knowledge it seeks to support? Is it low cost or at least good value for money? The impact on different school subjects through these contrasting approaches relates to the way in which conceptions of the subjects derive from where understandings of them are developed, whether in schools or in universities. For those in schools, little more may be done than enable teachers to work through commercial schemes as implementers of curriculum, much appreciated by those seeking to profit through the provision of such apparatus. For those following university-intensive courses, relatively low attention is given to the practical school aspects during the university element. In some countries, the approach has provoked some concern through its lack of connection to school practice.

5.5 An Empirical Study

I conducted empirical research in connection to recent changes in teacher education models. This research included a focus on a 1-year employment-based teacher education programme linked to my university but following the practices of a regional teacher education network comprising universities and associated schools (Brown 2018). The programme offered two routes that have transformed markedly during the course of the research, as a result of the models now being adopted in most schools. One route was for primary student teachers planning to teach mathematics as part of the broader primary curriculum to children aged 5–11. These student teachers would typically have studied mathematics at school until the age of 16 and later completed a university degree in any subject. All student teachers are required to pass a mathematics skills test administered by a government agency. The other route was for secondary student teachers specialising in teaching mathematics to students aged 11–18. These students would have completed a mathematically oriented degree. In the first few years of the research, in each of the two routes, the student teachers spend a total of 40 hours in university (e.g. a 5-hour day once a month for 8 months). The primary school student teachers spend about 6 hours of that total on the topic of teaching mathematics. The secondary school student teachers spend about 25 hours on the same topic. Yes those are annual totals on a 1-year course! For the rest of the programme, students work in paid positions full time in schools for the school year. The student teachers involved typically would have spent some time working in other jobs prior to teaching. They often chose the employment-based model as an alternative to university study. Many have expressed a preference for "wanting to learn on the job and receive a salary as they train" (DfE 2010, p. 23).

Our practitioner research on the programme, where I worked alongside the regular course teams, started some 10 years ago with an interest in the role of educational theory in this model of training. Theory had been part of university-based training in the past. Did it still exist as part of employment-based training? If so, what did it look like? Where was it located? In the early stages of the course, "theory" was often seen by student teachers as the stuff that was written in books and thus a bit distant from the immediacy of practice. Their priority was to get on with the job of teaching. Early experience in the first placement school was often very positive, but a number of the student teachers started to find working with just one mentor rather restrictive (see Jones and Straker 2006).

All student teachers moved to a second school 3 months into the course where they found that expectations and practices could be rather different. A new role began to emerge for the university component as the course progressed. Rather than focusing so much on what worked in a specific placement *school*, the issue was what worked for students more generally in *schools*. That is, the university sessions became redefined as venues where more generic teacher knowledge was created. Theory became the creation of analytical writing by the student teachers themselves, to support their practice across different schools. The university sessions initiated and responded to the student teachers' own classroom-based research as part of their getting to know how they might successfully work within a school classroom. They became a place in which their classroom practice could be critically evaluated against broader educational concerns.

For 4 successive academic years, on successive 1-year courses, the research team collected data through practitioner research methods. The longitudinal data collected within each year comprised examples of student reflections from regular recordings of university sessions, interviews, writing integral to course participation, assignments, correspondence between students and to tutors, reflective writing by the course team and interviews with students and with other staff responsible for mathematical content. Two extended interviews were held with tutors responsible for the mathematics element. Each session on each course included an element where progress was reviewed in terms of the changing ways in which the student teachers understood their professional challenge. These reviews incorporated regular reappraisals by the students of their own earlier writing as evidence of how they were changing.

Methodologically, a specific conception of "actor" was pursued within an action research model (Brown and Jones 2001). Research comprised active participation in wider cultural adjustments to new ways of being, in this case the move to different understandings of theory in new models of teacher education. A contemporary theory of the subject was introduced where the individual identifies with broader moves to new circumstances (e.g. Althusser 1971). These identifications produced changes in conceptions of the researched landscape and of the individual carrying out the

No idea what this means!

research. "Knowledge" here relates to a specific state of knowing that prevails in given circumstances. In that sense it is not universal. Yet, the imperative would be to constantly revise the narratives that guide our actions. Through living a story and becoming aware of its limitations, participants endeavoured to change to a new story. Or rather, endeavoured to keep the story of who "we" are ongoing and alive, as "we" adjust to ever-new conditions. Fail, but learn to fail better!

The third and fourth years of the data collection further included a specific focus on mathematics seen as a specific instance of our work on theory. In addition to individual interviews at later stages of the course in both years, nine secondary student teachers took part in a group meeting chaired by their university mathematics tutor, and eight primary students shared an extended discussion with the three tutors, which included the two course managers. These discussions were designed to review where the student teachers were up to in terms of their development as mathematics teachers on the programme and how the schools and the university had contributed variously to this. Analysis focused on how their understandings had changed. This involved sifting the interview transcripts to find instances of the student teachers' analytical connections to their teaching situations, such as evidence of their building an understanding what could work in schools generally rather than just in their current school placement. As we shall see, the earlier parts of the study reported on such shifts and specifically on how students looked back on their earlier reflections on theory. This was less possible in any detail with the mathematics focus as, given the course structure, the students were only in university for 8–10 days during their year's course with many demands being placed on their time. In the later interviews, the intention was to capture conceptions of classroom mathematics, empirically, as it was being understood by the students at later stages of the course but also through the recorded reflections of university staff either managing or teaching the course generally or specifically the mathematical elements. That is, echoing our work on theory, the research sought to avoid supposing that there was a correct version of school mathematics to which the teachers were supposed to subscribe. This chapter is guided by the more open research question: *How do student teachers discursively produce school mathematics?* In posing this question, there was an assumption that the student teachers could work on the ways in which they conceptualised mathematics towards revising these conceptualisations. In addition to data providing insight for the research team, the student teachers themselves looked at past writing to consider how their conceptualisations had changed. By better understanding their own past conceptions of mathematics, they would be able to move forwards.

Teachers then are *subject* to a specific models of teacher education, and our task here is to better understand how the assumptions implicit to the given model are articulated through the teacher accounts of their practices. This subjection restricts but also empowers the student teachers concerned. Individual and group interview data were analysed to assess the sources of influence or power referred to by the various parties and documents and cultural models governing conceptions of practice: inspection procedures, the school apparatus, the curriculum, the former governmental *Numeracy Framework* (or new school schemes or textbook choices), teacher education models, professional development initiatives, the parents, the children, etc. Transcripts of interviews and student work were examined in relation to how identifications with mathematics were understood (Bibby 2010; Solomon 2008). This analysis looked for evidence of how the mathematics curriculum was being progressively reconceptualised and re-characterised, in response to regulative apparatus (Brown, T. 2011), in relation to the wider curriculum (Alexander 1990) and to wider public conceptions of mathematics (e.g. Chap Sam 2012). The analysis sought to pinpoint how school-based training supported teacher subject knowledge. It further considered how university based teacher educators conceptualised changes to their earlier ambitions consequential to greatly reduced contact with the students.

5.6 General Findings

I now turn to how training for secondary and primary education produces the conceptions of school mathematics that govern teacher practice. The detailed government-produced "non-statutory" assessment framework for how the curriculum was to be covered had now been abandoned. Yet many schools still had schemes of work closely tailored to this framework. The schemes were typically staged according to the levels in the main curriculum. The student teachers, therefore, found themselves in schools where the curriculum structure was ever present in the shaping of classroom activity and of mathematics. Many student teachers felt coerced into teaching to the textbook or scheme of work. One secondary school mathematics student teacher described what she perceived as the relentless overseeing of the content and methodology of her teaching by her head of department: "The other day I was doing something a bit different and then he's going, 'You can't do the end of chapter tests on that because you haven't done exercise 5b!' I feel as though he wants me to do every single question in the textbook". Another extract from a discussion held with secondary school student mathematics teachers suggests that some freedom to apply the teacher's own ideas could be derived from following the school's scheme of work. However, this had to be assessed using the government Assessing Pupils' Progress (APP) framework, which the school followed: "I will plan my lesson, I use the scheme of work and I do this by myself. I don't have anyone to tell me what to do—no one checks that. There's no textbook to follow. I just teach my lessons so that they can do that, can use these words. At the end of topic, they have to do the APP".

Findings from the primary teachers demonstrated a similar exertion of school influence on what counted in mathematics. One student teacher, for example, in reflecting on a question posed about how he would decide to teach mathematics had this to say: "We have a policy, certainly for the four rules … I was doing ratio … and they were coming up with methods and I was looking at the class teacher asking, 'Shall we go down this route?'"

In English primary schools, mathematics is most usually taught with much whole class input, where the teacher must react to children's responses. This can be a risky business for student teachers when under the watchful eye of their mentors. In these situations, it was most important for student teachers to be seen to use the "correct" method. One student teacher described how moving from whole class teaching to individual activity, where children could experiment on their own ways of reaching solutions to mathematical problems, enabled him to "really see what the children could do". Ironically, he still needed to check the validity of the method used by a child with his mentor. She confirmed, with some hesitation due to the apparent deviation of the method from the more typical school approach, that if the children "got there, we'd probably support that [method]".

In short, we found that many student teachers learn to teach mathematics by participating in current school practices that closely follow the curriculum and the demands of national tests. Furthermore, schools and government agencies set criteria as to how this engagement was validated. Periodic national tests influenced the forms through which mathematical ideas were encountered. The consequence of these framings is that mathematics encountered in schools has a tendency to focus on those areas relating to the tests.

A significant aspect of the change in student teachers' understanding of mathematics related to how university mathematics teacher educators conceptualised their roles. They had been accustomed to spending a significant amount of time with student teachers in the university. Later, as increasing responsibility for training was relocated to schools, the content that had been previously covered in the university was condensed. The number of topics being covered was reduced, and those that remained were dealt with at a brisker pace. At first we, as university teacher educators, found this new arrangement quite stressful, compressed as our previous role now was into an increasingly small amount of time with the student teachers. Ironically, however, student teachers, thrust as they were into the hurly burly of school classroom activity, found the university sessions altogether more relaxing. Close pursuit of the curriculum in school framed their conceptions of mathematics, whilst university sessions provided reflective space. For primary school student teachers, the 6 hours at university early on in the course that provided a guide to the curriculum that they would be following were soon forgotten. Later in the course, mathematics was discussed as just one of the subjects that they were responsible for teaching. For secondary school student teachers, the 25 hours largely tackled issues relating to their teaching in schools.[2]

The orientation of the university component of the programme had shifted from one of input to one of response. Its role in supporting the student teachers had

[2] The time spent in university varied greatly between universities and had a lot to do with the market position of the specific university. Universities who needed to recruit more students were much more ready to abide by school demands to have students in school more of the time and to follow the school's preferences in terms of university input. Universities in stronger market positions were better able to insist on their preferred form of input but still within government limits that required most time be spent in schools.

relatively little to do with introducing broader issues in mathematics education, research in the field, subject knowledge being rethought as pedagogical content knowledge and so forth. More experienced university-based mathematics teacher educators found themselves subject to a very different conception of their practice to the one to which they had become accustomed earlier in their careers. The traditional content of mathematics teacher education, insofar as it was still addressed, was distributed across school and university settings. Many of the earlier ambitions advocated for teacher education (e.g. Askew 2008; Rowland 2012) or for subject knowledge (e.g. Ball et al. 2005; Davis and Simmt 2006) had been deleted from the list of training priorities. Askew and Venkat (2019) have carried out a recent review. If mathematics education research still influenced the practices of the student teachers, then the route through which this influence was achieved is not entirely apparent. It is also unclear how, within this model, one would seek to influence practice through mathematics education research. To whom would research about classroom practice be addressed and how would knowledge derived from this research filter into teacher knowledge?

5.7 Student Teacher Experience of School Mathematics: Some Data

This section provides data on the mathematical aspects of the teacher education process. Secondary and primary student teachers are looked at in turn with a view to highlighting how mathematics and its teaching are variously framed within the conceptions of their own professional practice in this area. In both cases the research strategies doubled as attempts to encourage the students to describe the worlds of their teaching, which so often would have been relatively private. The descriptions were seen in terms of making sense of their practice towards transforming that practice.

5.7.1 Secondary Student Mathematics Teachers and University Tutors

Issues relating to the university element were initially seen as peripheral ("*Reflective Account Two*? Whatever! It'll get done") or disdain ("It's paperwork...I hate X"). The dominant theme in discussions was the immediacy of practice ("The teaching's going fine - if I could just focus on that, it would be ok!"). It becomes clear, however, that the students feel that the teaching is not always "fine". In significant ways, it is not fine and the discussions sought to dig deeper.

There was much talk about the vagaries of the assessment of the mathematics curriculum in relation to the performance of those taught. It is now increasingly common for those pupils taking public 16+ examinations to be entered early.

Obtaining a prized pass at grade C[3] at this level was seen by some student teachers as introducing significant problems in subsequent pupil motivation and knowledge levels in the remainder of compulsory schooling in the subject:

> I've got the most bizarre class, a top set [16+] who have all passed [at C and above] and who've all got a different history. … [They] don't want to pay any attention at all to what's going on unless it's directly relevant to them. The theory is that they are an improvement class, trying to better their grade, so it's been really tough. I think it's a natural consequence of early entry and promising them all if they pass early, then they don't have to worry about maths anymore. Some of them have done it … purely on common sense and ability, in my view. They've turned up, done no work and got a C on the paper because it's pretty easy - don't know any of the higher [level] content [included within the exam] and don't want to know it. Others have managed a B or an A … and covered a lot of it – got one or two gaps in order to improve. Deciding what strategy to do with them has been really tough. You can't do thirty different lessons can you?

Attempts by a university mathematics tutor to explain pupil behaviour as symptomatic of an assessment system driven by performance, rather than the intrinsic worth of learning, were not, at first, readily taken up by the students. That some pupils were differently motivated was acknowledged, however. The students themselves appeared to reduce the level of challenge that they faced personally in mathematics. One saw it as a need to "going back and remembering things", to reaching a solution to a pupil's (and their) immediate problem, rather than any inherent lack of understanding about teaching and learning on their part and a need to develop this. In arriving at a "quick fix" to the challenge faced in their learning, and nothing more, the students' behaviour seemed to mirror the behaviour they witnessed in their own pupils:

> 1/3rd into 1/5ths? I don't understand it numerically - I *can* do it.

> I'm challenged …whenever I teach [post 16+]. I'm always there and they're going, 'So is this right?' and I'm like, 'Ermmm - I'll just get a bit of paper'. But I try and do them beforehand, if I've got time, you know, work out all the answers myself and then I've got my work and I can go, 'Hang on a minute. Yeah, that's right'.

> I think it's a question of refreshing your memory sometimes. I've got histograms tomorrow and I think, 'How do these work?' And you just go through and … I remember.
> I find the [statistics] hard. When I was doing the [16+] stats, I thought, 'I'm going to have to teach myself how to do this.'

> [Vectors]…they're *my* nemesis!

[3] Grade C is widely recognised as the pass grade in national 16+ examinations. Occasional footnotes will act as a guide to time- and context-specific terminology, but the point being made is that it is very complex to the extent that it rarely achieves substantial meaning. An earlier version of this chapter appeared in an Australian journal where demands for clarity were hard to meet in the straightforward way that might have been hoped for. In an earlier study (Brown and McNamara 2011), we interviewed student teachers over 4 successive years of a teacher training course where students in each year were asked the question "What is mathematics?" It was intriguing that students in successive years gave answers in which the mathematical ideas were cladded in ever denser locally specific administrative descriptions.

A second area of difficulty for the students related to a pragmatic and superficial approach of getting the mathematics lesson done, rather than teaching for learning. The lesson was easier when explicit teaching did not have to take place:

> I find it harder at the lower end…bottom set Yr 7 (11+). How do I know how to write pounds and pence…so much of it seems instinctive…I find that end more difficult. You can take it [understanding] for granted [with the top sets]. [The bottom sets] question it more – the top sets are just in kind of, in the mode of, we learn the method and do things. We do it for the exam, like little robots - quite happy. Whereas the bottom sets can't do it that way. They want to know why it is and they don't understand what's going on and they're mixing up different things they can remember. Some [pupils] just understand it without you delving into it. Some [pupils] discover it for themselves …and some don't and they're the ones who get it wrong and that's why they don't get it. Even if you try to drill them, because they don't understand it, they're not going to remember it…what about the ones who've never discovered it? We teach those that already understand it and knew it and they practice it, and they do well. And I think my challenge is how to move some [pupils] on who didn't understand it first time, who haven't got their head around it. How do you move them on?

Some of the student teachers recalled helpful materials issued by the university, which delved into such topics as pupil misconceptions, strategy games and "scripted lessons". However, one of them talked of being "swamped by other (training) agendas" as an excuse for not referring to the materials as much as she would have liked. Now spurred to "get underneath what the maths is about", during a group interview, a tutor asked some students how they decided what mathematics they would teach and how they would teach it. In responding, the students became very animated. The slavish adherence to textbooks was contrasted with the supposed liberty of following a scheme of work. There was little to support them in either in their quest to teach mathematics in ways that might encourage interest and understanding:

> At my school, it's just a textbook basically you're working through the textbook and do X number of chapters per half term. My head of department is really hung up on it. … Literally you follow page after page …and you just did it in the order of the textbook … These are our schemes of work written up by the head of department for Yr 7 (11+). It tells us what topic we are doing, when …what they should be doing, what are the key words. We sign up to an APP [*Assessing Pupils' Progress*] programme,[4] which we can use if we want to … All our kids will start a lesson with this. They'll identify stuff they can already do …what they have to do to get to [National Curriculum] level 5.[5] I will plan my lesson, I use the scheme of work and I do this by myself. I don't have anyone to tell me what to do - no one checks that. There's no textbook to follow. I just teach my lessons so that they can do that, can use these words. At the end of topic, they have to do the APP at the end of that.

---

[4] *Assessing Pupils' Progress* (APP), administered by the government's *Qualifications and Curriculum Development Agency*, has been developed for optional use in schools in England and Wales to enable teachers to assess pupils' work consistently across both the secondary and primary National Curriculum. Many schools have abandoned this scheme as a result of its excessive demands.

[5] Pupils would start their primary schooling at level 1 and transfer to secondary school at 11+ where the average level would be 4 but where pupils would be spread over a range typically between levels 3 and 6. Formal tests take place for children aged 7+ and 10+ where the later test results are published. There are informal tests at the end of each primary school year referenced to National Curriculum levels.

5.7.2 *Primary Mathematics Student Teachers and University Tutors*

Towards the close of a group interview, some primary student teachers were prompted about the extent to which assessment was an issue in their development as teachers. Unlike their secondary peers, they had neither introduced nor, as we are to see perhaps, spoken about a dilemma concerning performativity in tests and external assessments, at least not on the surface. Some more persistent primary voices showed, however, their developing sense of skill in assessment practice and the multiple filters through which it needed to be understood:

> I find assessment quite difficult sometimes … For example, if I'm doing "direction" with the lower ability [pupils] that might be my [National Curriculum level] 2bs or 2cs. They just need to know a quarter turn and a half turn, whereas the higher ability need to know quarter, half-turns and three-quarter turns as well as clockwise and anticlockwise. … Sometimes I'll come to the assessment sheet and there'll be nothing in there for whatever it might be and that's when I get flummoxed with it… Am I doing the right thing here? … Sometimes you won't find it … It just won't be there … I never really thought that in first half of the year. I just was differentiating because I knew 'that was harder'.

At the group meeting, the primary student teachers were pressed directly about whether performance identified in the secondary discussion was indeed an issue for them. About a third of the group talked about overhearing the Year 6 (10+) colleagues in conversation about things "coming up on the test". Tests were held for pupils at the end of their primary schooling. Some felt that assessment was much more relaxed for the learners in a primary setting with no real "pushing" of performance. Seven year olds often remained unaware of the interim tests that they completed. Others felt that the extent of accountability was dependent on the ethos of the specific school. In one school, it was normal to maximise levels of performance, "as soon as a piece of work was finished, wham, it was levelled" [according to National Curriculum level]. It was enough to maintain a standard in a second school. Most student teachers recognised a key difference between their own experiences of assessment and that of the secondary student teachers. For a higher proportion of primary student teachers, the presence of National Curriculum levels was a continual process, formative rather than summative, as was the case for most secondary student teachers with their focus on tests and exams. The primary students agreed that this led to an ongoing pressure to monitor progress and not simply react at the end of the year. Nevertheless, as the discussion continued, it gradually became clear that these students were developing an awareness of the spectre of accountability haunting their teaching. Mention was made of the "expected 2 sub-levels of progress" (e.g. moving from National Curriculum Level 2c to 2a) learners were expected to make in the year and an awareness that if this was not the case, "you're (the student) going to be questioned". The students voiced their growing concerns about perceived lack of progress, "why is this cohort not scoring is constantly in your mind" others spoke of the impact of children being inaccurately assessed by colleagues, claiming, "it'll look as though I've taken them backwards".

Interestingly, one primary student teacher wanted to address children's understanding of concepts and distanced himself from governmental expectations graded as successive "points", which he saw as unattainable for children in his setting. This extract is revelatory of just how immersed student teachers become immersed in the regulative discourses that define their practice:

> They're expected to get Point 9[6] by the end of [4+]. We have kids who are on Point 2 or 1. They're not going to get to a Point 9 and if we have two or three children on Point 9 at the end of the year then that's average. … We have interventions … in place for the highers [higher ability pupils] and the middles and the lowers …There's only so many of us…90 kids and three teachers … It's often the Teaching Assistants[7] who deliver the interventions and they often haven't had the training… It is in our interest to raise our understanding and keep this in mind but I almost think it's an impossible job, impossible to get to expectations… If [government inspectors come, they are] not interested in why. They wouldn't take that into account.

Reflecting on the primary student teachers' discussion, one of the tutors made the following supporting observation:

> They are not making the connection between the children's very closely targeted learning and the assessment processes that are informing and driving this quest. Levels and targets have just become part of their professional dialogue. They are not asking what makes a child [National Curriculum] Level 2a and how the teacher knows that it is reasonable for that learner to have progressed to Level 2c by the end of the year. They operate currently by planning lessons that allow children to progress with their individual targets without knowing clearly where these came from or where they go to, just that's what they are required to do. So, like the goldfish in the bowl being unaware of the water, they are unaware of the assessment driven process. It just is.

That is, the student teachers are not always aware of how the regulative discourses were *shaping* their practices. As with their secondary peers, however, primary students were very aware of the policies and associated apparatus *validating* their practice. This was more vivid when the student teachers talked about applying the teaching methods preferred by the school or those featured in the *Framework*, such as in following calculation methods different to those encountered by the students in their own schooling:

> Putting myself in the place of the child is difficult. … The way that I would work it out is slightly different. … I am having to constantly address my own way of dealing with these problems. We have a policy, certainly for the four rules [of arithmetic].

> I was doing ratio… and they were coming up with methods and I was looking at the class teacher asking, 'shall we go down this route?'

Clearly some students felt constrained about following the children's line of enquiry for fear of wasting time, or far worse, confusing learners by moving away from agreed models. However, mathematics was obviously a subject where

[6] This refers to an assessment tool used with children aged from 3 to 5 years. It is pretty meaningless to this author as well!

[7] Teaching Assistants are commonly employed in English schools alongside the main classroom teacher.

they had to actively stand at the front of the class and teach, rather than simply respond to students' individual work:

> That would be my main teaching I'd say here's one method of long division or accounting method, here's one way of doing it, did everyone get the same answer did anyone get it in a different way.

> I've taught [primary maths] and never seen children working out of books and teachers responding. There's an oral starter then shared input, paired or group work then independent work. And it's that shared input is the essential bit for you ... to see the differentiation in the class if it's working or if someone needs to move group ... Luckily in the first term I was in the Maths co-ordinator's[8] class. She would use three different types of input... to meet the needs of different learners... It was amazing to watch, very hard to do.

Some students, however, did describe points where children were deemed to have reached a point where they could choose their own mathematical processes in an assessment activity where they were told "If you want to use the number line ... use which methods you want to ... there were no restrictions ... which really helped me to see what the children could do". There was an emphasis on the *how* of teaching in each phase. The school scheme was a key part of this. There was, however, room for flexibility depending on the needs of learners, teaching and learning policy. Calculation and method played much larger roles in the regulation of primary students' working practice.

5.8 Discussion of Data

The above descriptions give some insight into the varied ways in which the student teachers map out the territory of their practice. How do they talk about the world that they inhabit? Which points of reference are mentioned most? How are those points of reference probed within the research orientation of the course? The scene depicted is dominated by an ever-present culture of assessment. The teaching of secondary students was a step-by-step targeting of 16+ ambitions. Primary students followed textbooks and schemes where the assessment levels were built into the "goldfish bowl" of their practice. I have spared the international reader much of the frequently changing jargon, of "levels", "key stages" and "points" that dominated student accounts to avoid those details from distracting attention away from our more general concerns. Yet the terminology did much to partition mathematics according to discrete learning objectives and local discursive preferences. Mathematics is defined by alignment with a criterion-referenced listing shaped by the demands of this assessment. Meanwhile, the university element had become quite restrictive in its very brevity, very much so in some universities. Familiar features in many models of mathematics teacher education had become marginalised through demands for compliance with current practices in schools. In the reality of

[8] Mathematics coordinators are teachers in primary schools overseeing mathematics teaching throughout the school.

the training experience for many teachers in England, explicit space for developing the intellectual dimensions of practice has become much reduced. The teacher's conception of his or her own professional identity is tightly referenced to the regulative structure set out in policy documents. Success in teaching her was referenced to the then current model in schools whose achievements had so dissatisfied the government prior to the reforms.

To summarise key issues raised in the data, one might highlight:

5.8.1 *Performance-Driven Assessment Affects the Nature of Subject Knowledge*

School-based practice has been driven by the need to meet assessment requirements. Trainee teachers are given fewer opportunities to conceptualise other modes of practice. By emphasising the elements that are more likely to be tested, subject knowledge may be diminished. Current conceptions of school mathematics and science, for example, are supported but only in a narrow way if judged primarily by their ease of assessment. Less emphasis is placed on pupils being able to adjust to future demands. This emphasis drives compliance to external demands in which student teachers and their pupils play a smaller part in the construction of the subjects. There is a culture of "getting it done" or "giving the method" rather than teaching for understanding: "Does that make sense… is that realistic?" An occasional decision to "step back" from the formal in the name of building understanding, "light bulbs were going on everywhere", was an exception rather than the norm in the anecdotal material. The thrust in English schools over recent years has been towards supporting skills-based agenda. For example, as mentioned, following a governmental-led "back to basics" campaign, England improved its position in mathematics on TIMSS in 2007 whilst dropping in its rankings on more problem-focused measures within PISA in 2006. This led to new complaints about England dropping its standards with selective reporting by both newspapers and the government. Being a teacher is understood in terms of shaping subject knowledge in line with curriculum specification to meet the required forms to suit the given class composition. This external specification can lead to some issues of continuity in education in England where successive phases (e.g. exams at 16+, 18+ and university degree level mathematics, in England) each work to a different discursive frame as to how teachers, students and mathematics are each understood.

5.8.2 *School-Based Training Can Nurture Narrow Administrative Conceptions of Teaching*

For student teachers on school-based routes, being initiated into teaching by way of their placement schools' insistence on following specific textbooks "page after page" in some instances diverted student teachers from trying out ideas introduced in university sessions. Taking the case of mathematics: This narrow approach is perhaps unsurprising for primary students (those teaching ages 5–11) who have usually not gone beyond their 16+ examinations in mathematics or science. This narrow approach is perhaps unsurprising for primary students (those teaching ages 5–11) who have usually not gone beyond their 16+ examinations in mathematics or science. Yet even for secondary teachers (those teaching ages 11–16) with formal mathematics backgrounds, there was some trepidation in relation to the mathematical demands of teaching. Many student teachers in mathematics and science now feel the need to follow special courses to enhance their subject content knowledge in advance of commencing formal teacher education.

Yet these occasionally negative assessments of school-based training limiting the development of subject knowledge are countered by some additional pedagogical factors relating to a stronger school role in teacher education that conferred some benefit.

5.8.3 *Practice-Centred Learning Can Improve Participation in Schools*

Some school practitioners see virtue in employment-based models because of their immediate concern with the demands of the classroom. Mathematics is a function of its location. A mentor responsible for overseeing such students in a demanding inner city location spoke of how the school's greater input allowed more investment of support time aimed at enabling new teachers to survive and function in difficult circumstances (cf. Clandinin et al. 2015). For a school with a well-developed scheme of work, student teachers and pupils alike may benefit from the student working to a clearly defined structure as a shared enterprise with colleagues. Such a *community of practice* (Wenger 2000) may supply genuine opportunity for students to experience an insider perspective on being a teacher. As one student put it: "the behaviour of the students is challenging, but we're encouraged to take risks and try out activities". Some school-based students were offered jobs by their placement schools prior to the course being completed. This was good for the school to have found a suitable teacher in an area of persistent teacher supply issues but could reduce the student's motivation to exceed the already limited academic demands.

5.8.4 *The Enforcement of a Centralised Curriculum Supports a Collective Vision of Learning*

The motivation behind the somewhat insistent centralised curriculum was centred in administrating the many teachers who lacked adequate subject content knowledge and professional capabilities to work without explicit support towards a collective set of ambitions. Many student involved in training to be secondary mathematics teachers do not have the requisite mathematics-oriented degree. Any collective arrangement requires compromise, and unnecessary guidance to those teachers who were adequately skilled was seemingly a low price to pay for wider participation in a shared arrangement. Education research is sometimes predicated on finding more refined pedagogical strategies for a teacher to follow whilst neglecting the reality of teacher recruitment in terms of individual skill, where prototypical secondary mathematics teachers were in the minority. Alternatively, student teachers might creatively identify with approaches spanning a larger population of teachers as a mode of support for those with lower confidence or different specialist background.

5.8.5 *Research Is Directly Focused on Developing Practice*

Many instances of education research are finely tuned on issues unlikely to be encountered in preservice training courses. Within school-based models, however, the students themselves may have the opportunity to participate in forms of practitioner-oriented research made possible by the immediacy of ongoing school practice (e.g. Hanley and Brown 2016, 2017). The university element that had often been irrelevant for many students in the first instance can later become an effective critical platform for inspecting and reflecting on their own school practices. This platform potentially provides an opportunity to articulate the shaping of practice from an alternative location in which everyday demands could be understood against a wider context. Rather than thinking what would work in the current placement *school*, the concern became that of thinking more broadly about what would work for them across *schools* more generally. So rather than student teachers being subservient to a map dictating the format of their practices, they had some influence over how the map was created and how it guided their generic practices as a teacher. These opportunities to connect school with university input featured less in the Bologna Process prevalent in most European countries since university and school phases are sequential.

How do we know this?

5.9 Conclusion

As seen, school mathematics is a function of the educational domain in which it is encountered and hence of the discourse that characterises that domain. That discursive structure can shape the actions of those subject to it, yet it may be possible to step outside. This chapter has documented some instances of mathematics teaching practice resulting from modified conceptions of teacher education that are emerging in a number of countries. The teachers' conceptions of mathematics developed without a great deal of explicit instruction from university specialists in the area. Rather, the teacher education function was achieved through the student teachers being immersed like apprentices in the infrastructure of schooling and learning to speak the local languages. In the approach described, the student teachers were primarily guided by their school mentors through centralised curriculum documentation or by textbooks chosen by head teachers. That is, the students' mathematical pedagogical knowledge is derived from their own practice referenced to existing or required conceptions of mathematical knowledge and patterned on the associated apparatus. Their way of talking about mathematics teaching mirrored the official discourse. Consequently, there was a strong reproductive dimension to the student teachers' understanding of school mathematics and its values. Mathematics is defined within very tight boundaries that give it little space to be something else, such as mathematical constructions generated by the teachers or pupils themselves.

Specifically, in the data presented, mathematics derives from different types of encounter in a model of teacher education.

On the one hand, mathematics was understood in terms of fixed results, levels and following procedures. Little opportunity was provided for the student teacher to develop an autonomous professional attitude to the generation of mathematics in the classroom. Rather, the students were *subject* to an externally imposed curriculum as represented by the mentors to whom they were assigned. They understood their own professionalism and identities in those terms. The "goldfish bowl" of practice denied space to a more externally critical attitude in favour of training through immersion in school. Although there had been some stepping back from the more prescriptive aspects of the curriculum guidance, the student teachers are still subject to a legacy in which conceptions of teacher have little room for manoeuvre, predicated as these conceptions are on specific constructions of mathematics. Some students, however, feel more secure with these arrangements in an area where they may lack confidence. Their own mathematics background may also have been centred on test performance rather than on understanding limiting their capacity to step away from pre-defined pedagogical routes. Such students needed to know the topic in advance as defined by the book or scheme rather than treat the encounter as a process of shared learning.

On the other hand, the new model does provide an avenue through which student teachers and their tutors can experience the teaching of mathematics from new angles. This dimension however is at risk as more teacher education is scheduled to take place outside of university settings. In the model described, student teachers retained some possibility of inspecting their practices in school from an external site so that their insider experience of meeting immediate demands can be reviewed against a more holistic understanding of what they are trying to achieve. University tutors, meanwhile, provided a responsive role in helping students to confront demanding classroom challenges in more creative ways, albeit in terms of administering mathematics to the prevailing model.

There is another factor that has become more prominent since the empirical research described here was carried out. In England this is called the National Student Survey. This survey provides students with the opportunity to evaluate their university tutors, rather like the "Rate my professor" site more commonly referred to in the United States. The National Student Survey has become a powerful instrument in regulating university teacher education practice, where university staff find an increasing pressure to be responsive to student demands in connection to styles of teaching they receive. This restriction on teaching style further undercuts the prioritisation of research-led teacher education practice. My colleague Jonas Thiel carried out an extensive survey of how this infrastructure functioned as regulative apparatus concerned with surveillance, with reference to the work of Foucault and Barad (Thiel 2018a, b, 2019). Such apparatuses might also be, for example, quotas, nominations, accreditations, qualifications, TIMSS, PISA, financial management of teacher education, teacher educators making sense of their practice situation, university assessment/management of its employees, universities specifying job descriptions or recruitment procedures. Ofsted and National Student Survey grading systems, similarly, arbitrarily impose certain values to effect specific distributions of teacher education across providers and shape the human actors involved. "Apparatuses are themselves material-discursive phenomena, materializing in intra-action with other material-discursive apparatuses", … *where apparatuses are not mere observing instruments but boundary drawing practice – specific (re)configurations of the world- which come to matter* (Barad, emphasis in original pp. 203, 206). We cannot "raise" standards but only reconfigure what they are and thus change the way in which those people adhering to standards are noticed or understood. Fears are emerging that such surveillance is becoming an intrinsic dimension of technological development across the world. Griffiths (2019) and Strittmatter (2019) both discuss the case of China.

This chapter has focused on specific themes pertinent to the situation in England where school-based training has become legislated as the norm (Brown 2018). As seen, the government has indicated its preference for expanding this type of provision yet further. Indeed, school-based teacher education can be developed to provide supported participation in communities of practice where mathematics and its teaching are built as more collective enterprises shaped around the needs of mainstream schooling arrangements. This however would be an unpopular move in some quarters. The students' conceptions of mathematics and its teaching on the course described are crafted around the apparatus of administrative control, which are restrictive, expressed in terms of curricula compliance, or fitting in with existing school practices. This administrative restrictiveness in the name of policy implementation is potentially counter both to pupils achieving a positive disposition towards mathematics and functionality in the subject in later study or professional life (see Pampaka, et al. op cit.). These conceptions also diminish the teacher's professional life, reduced as they are to following someone else's model during their formative years as a teacher, where experience across different placement schools is uneven.

In the model described, research carried out by student teachers fuelled a more generative attitude to practice that could be supported and developed in university sessions. That is, a practitioner-oriented reflective approach comprised an integral dimension to practice in school and the university sessions. Here, research is not seen as knowledge confirming a desired state of affairs in the manner of yet more insistent external demand. In the approach described, the university, rather than being the font of knowledge depicting models of good practice, provided a critical platform from where analytical apparatus could be created to support the generation of knowledge in developing practice, to counter excessive compliance with those external demands. The demands may shape our practice, but perhaps we can develop the capacity to distance ourselves from the discursive parameters that deliver those demands (see, Brown et al. 2019).

"Subjection consists precisely in this fundamental dependency on a discourse we never chose but that, paradoxically, initiates and sustains our agency" (Butler 1997, p. 2). "Power not only *acts on* a subject but …*enacts* the subject into being" (p. 6, her emphasis). That is, the discursive arrangements that define practice can be inspected from outside and then turned against themselves to provide leverage into a new space. For Butler (1997, 2005), the very restrictive positioning as subjects creates a framework for resistance. "For what is it that enables a purposive and significant reconfiguration of cultural and political relations, if not a relation that can be turned against itself, reworked and resisted" (quoted by Davies 2006, p. 425). The more marginalised role for student teachers and their tutors can be re-crafted as a critical platform from which both tutors and trainees can inspect the stories governing their respective practices and the opportunities those stories provide for the development of analytical apparatus.

In a later book (Brown et al. 2019), I have explored possible changes to practice in the light of the empirical study described above. That book had a specific focus on teacher education in the case of secondary English education where student teachers were encouraged to narrate their own path into teaching. They sought to track their transition from being English undergraduates engaged in nuanced discussion of Mrs Dalloway's troublesome day to, 18 months later, confronting real-life 14-year-olds with an attitude problem towards phonics. In keynote lectures to subsequent cohorts of students, I was able to show examples of the personal pathways followed by previous students through their own reflective analysis of their progress through the school-based course into full-time teaching. In an associated paper, we put it thus:

> Working in a Lacanian theoretical perspective, we encouraged students to remain attentive to how desires or wishes influenced their perceptions. In particular, students were tasked with noticing how projected fantasies dictated a sense of what was possible and how language might be used to frame things differently. Students faced difficult choices. If they decided to stick with current interpretations, to suture meaning here and not there (Žižek 1989) what developmental opportunities were being missed? There can be significant risk in a speculative process of inquiry whose outcomes are not guaranteed in advance. Students were asked to remain sensitive to how the desire for certainty influenced narratives of 'what really happened', and how these might be further analysed. (Hanley and Brown 2016, p. 15, see also 2017).

Meanwhile, Chapman (2019) has conducted her own narrative-centred research in mathematics education. Perhaps the new role of universities is to provide a platform from where both tutors and trainees can critically analyse the issues arising in school practice. This new focus would be on building generic analytical capability that supports learning by the trainees in association with their school-based mentors. The challenge would entail supporting trainees in becoming more independent research-active teachers through building a productive critical relationship between university sessions and their developing practice in school. Here universities would assist trainees in developing practitioner-oriented research and connecting it with the broader body of research knowledge. That is, reflective practice would comprise a creative ongoing process of practitioner research that progressively defined the parameters of teaching whilst negotiating a path through the external demands that trainees will surely encounter. Collaborative, reflexive, practitioner-oriented action research would underpin successive reconceptualisations of practice towards enhancing trainees' abilities to claim intellectual space in these regulative times. New priorities have shifted teacher education towards schools and may require aspirant teacher educators to remain in schools or to change their practice to meet the new demands.

Ultimately, conceptions of improvement are very much a function of the country, or even local community, in which they apply and the state of affairs prevalent there. And it is this sense of contingency that underpins this chapter's focus on adjustments to new paradigms. In particular, it is unhelpful to suppose that we could identify trajectories of improvement that apply across all people and all phases of development:

> Time metaphors abound in the hegemony of educational discourses seeking "improvement" or "progress", [and in England,] towards "greater effectiveness" or even the dizzy heights "outstanding status" or "world leading", thereby sublimely producing standardised modern notions of change, orientation and a correct way forward. But the reflexivity of life can result in us celebrating and protecting our current diversity rather than nurturing futures that might not allow the new to happen. Emancipation, for us, is about enabling a critique of the discursive platitudes that have locked our resolutions into overly familiar pathways. (Sant and Brown 2020)

School subject knowledge has come to be a function of this newly described world, backed up by governments using these conceptions to set their policies. There is always a cost in the form of suppressions resulting from such generalist suppositions. To represent mathematics as universal, spanning nations and generations, comes at a price in terms of teachers' ability to identify with the modes of education privileged in such comparisons.

Chapter 6
The Point of My Own Teaching

The previous chapter has discussed ways in which teacher education practice has changed in my own country such that student teachers spend most of their training period in school. The majority of teachers do a 1-year postgraduate course following a degree in their specialist subject, where there may be no pedagogically oriented focus. For mathematics secondary specialists, the time spent in many universities during that year on specifically mathematical themes is around 25 hours, and for those teaching primary mathematics as one of nine curriculum subjects, it is around 6 hours (Brown 2018). More prestigious universities with strong market positions can insist that schools follow their preferred approach, but there is still a need to comply with government requirements that specify the minimum number of hours to be spent in school during the training year. The majority of students entering teaching in England do so through universities more susceptible to market pressure centred on student recruitment where the influence of schools is much greater. In such settings schools can be stronger in setting the terms of the partnership arrangements, where universities and their students are seen as providing support to schools. The point for the purposes of this book is that mathematics education research does not necessarily have an opportunity to present its recommendations in such arrangements. Classroom practice in mathematics, certainly at primary level, more usually follows commercial schemes. The agency of individual teachers and of many teacher educators to make autonomous professional decisions is held in check. Inevitably these restrictions place limits on what mathematics can become in many schools.

T. Brown, *A Contemporary Theory of Mathematics Education Research*,
https://doi.org/10.1007/978-3-030-55100-1_6

6.1 Spatial Apprehension[1]

Here I will describe some work that I have done with some students where I have explored my role as a mathematics teacher educator in situations where more opportunity to experiment is enabled. This follows up and develops discussion of some activities that I described in my book *Mathematics Education and Subjectivity* (Brown 2011). On this occasion, I provide some critical analysis carried out by students of their own work on various mathematical activities. I have a regular weekly class with a group of adults, typically in their 20s and 30s, from a range of professions, retraining to be mathematics teachers in British secondary schools. As non-mathematics specialists, they are offered the opportunity to develop their understanding of mathematics over a period of 6–9 months prior to entering the year-long, school-based postgraduate "training" course now typical in England and Wales. Such university Mathematics Enhancement Courses (MECs) offer a way to develop greater confidence in mathematics, seen from the perspective of how it might be taught in schools. For many people embarking on these courses, their own schooling was characterised by a mode of teaching centred on getting through exams. On some teacher education programmes, the mathematics input is so brief such a view of teaching might be difficult to shift. MECs can provide a luxury resort where more creativity can be possible revealing the delights of mathematics in an exam-free zone prior to the frenetic schedule of a teacher education course. MECs might then not so much be seen as basic training for future course survival but rather a vision for a more exciting career beyond, something to hold on to during the sometimes dark days of intensive training towards formal accreditation, as described in the previous chapter. Most of the units comprise straight mathematical content. The unit in the MEC featured in this chapter and the next is more designed to see the learning of mathematics as *research*, centred on the question: "What is it for someone to learn mathematics?" From day one of the course, the challenge is to pay attention to how we learn and how those around us in the session learn, with a view to developing a way of describing how the children to be encountered in school learn. Each of the 20 3-hour sessions comprises a mathematical activity to be explored with view to articulating how learning happens for those involved. The range of activities is broad to ensure that mathematics is experienced from many different perspectives and to resist the common expectation that school mathematics is just about doing exercises. My ambition as their teacher is to introduce them to a broad range of mathematical experiences prior to the intensity of the subsequent training year where a relatively reductive version of school mathematics will be encountered.

---

[1] This section draws on material first published as:

Brown, T. (2016). Rationality and belief in learning mathematics. *Educational Studies in Mathematics*. *92*(1), 75–90 and

Ballantyne, M., & Brown, T. (2016). Close encounters of the curved kind. *Mathematics Teaching 253*, 28–32.

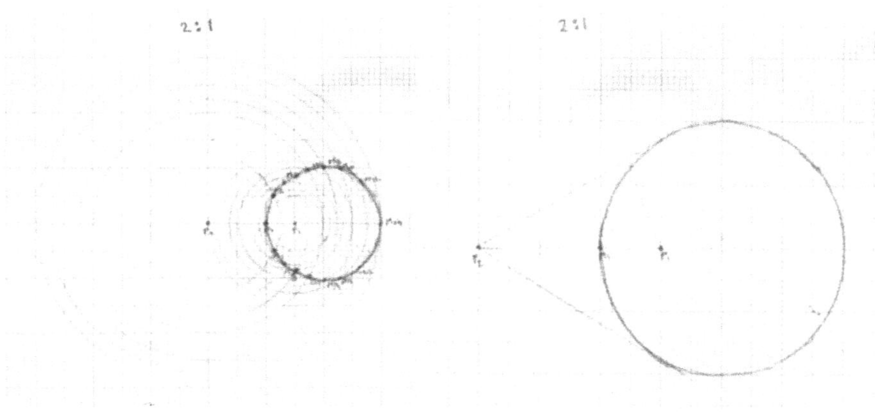

Fig. 6.1 Constructions of circles

In some activities directed towards enabling the students to develop their spatial awareness as a prelude to a more formalised approach to geometry, they were invited to explore various body movement activities. My intention here was to explore geometric entities from multiple perspectives, especially from inside very large versions of these configurations. In an example, a student was asked to position herself between two dots on the ground, which were about four metres apart, but where she was twice as far from one dot as she was from the other. She was asked to walk so that she was always twice as far from one dot as she was from the other. Alternative interpretations were provided as various students attempted this challenge. My hope was that there would be a lot of variety in the responses so that alternative ways of making sense of the situation could be shared and compared later. A few students produced drawings linking the points that satisfied the conditions, showing that they made a circular path (Fig. 6.1):

Another student reported a completely different experience in connection to the same problem. His response was to produce the following set of equations, seeing the same circular loci but in algebraic terms:

$$(x-12)^2+(y-0)^2=6^2$$

$$x^2-24x+144+y^2=36$$

$$x^2-24x+108+y^2=0$$

I checked that the formula is correct by solving when $y = 0$.

$$x^2-24x+108=0, x=6 \text{ or } 18$$

The issues became yet more complicated as the problem shifted to remaining twice as far from one dot as from the other in three-dimensional space. The challenge

Fig. 6.2 "Conceptualising" a three-dimensional object using string

provoked much ostensive gesturing alluding to points beyond immediate grasp and constructions out of string to confirm speculations (see Fig. 6.2).

This experience was written up at home as part of a diary package that would eventually be submitted for the course assessment. The course is set up at the outset as a place where we all research how people learn mathematics. Starting with ourselves and our own learning, we tell stories of our experiences towards building some sensitivity to understanding how similar situations can be experienced in very different ways and that our own learning can be enhanced by trying to see my problem through someone else's eyes. We explore our respective beliefs and expound the rationalities that link them. For example, two students in the same subgroup experiencing the same discussion documented the different ways in which they saw their colleagues had made sense of the problem:

> people do not visualise the same problem in the same way … each individual gave very different, but equally valid, explanations … for seeing a circle in 3D …: a penny being spun around at the end of a piece of string; modelling the shape with your hands; imagining being the origin of the circle (therefore being inside the shape) and what it would look like looking in each direction; imagining the shape being built up from the established points which were on the ground.

This student produced a drawing to show her own image (See Fig. 6.3).

> What was interesting was the different ways in which we described our thoughts and showed them to the group. N was thinking and demonstrating as if she was inside the shape.

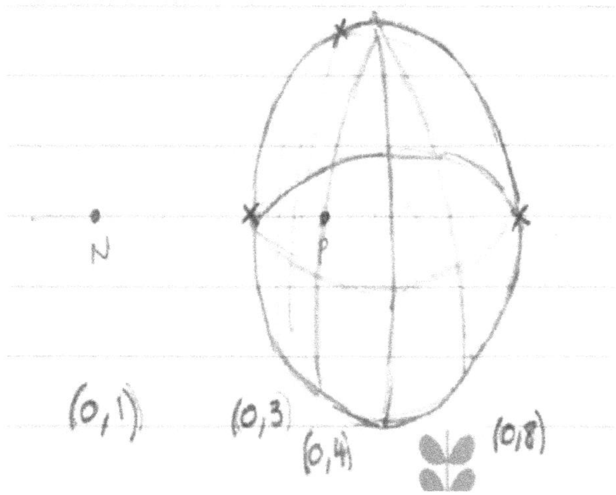

Fig. 6.3 A sphere

> S looked like she was thinking outside the shape. I thought of the shape spinning on a fixed axis, this fixed axis being along the line of the two fixed points; I was also visualising the shape on a 3D-type computer programme. J thought of it spinning on a fixed axis and used the idea of spinning of a coin to explain. We all agreed that the shape made would be a sphere.

This second student continues:

> Even though we all had different conceptualisations of what the curve looked like in 3D we could agree that we were talking about the same curve. This happened because each individual explained their visualisation and it fitted in with each of the other individuals' conceptualisation. For example, I imagined looking at the shape from the side seeing the established points and building it up from there. However, I could see that the image of a penny being spun fitted with my visualisation so could assume that person was having an equivalent (and yet different) visualisation to me.

So, a sphere is a sphere, but different people can experience it in different ways at different times (as a set of equations, drawing, as an imagined ostensible object, like this ball). The attempt by a student to reach out to someone else's description often required the intellectual challenge of comparing it with their own vision. "Every interpretation is partial, 'embedded' in an interpreter's ultimately contingent subjective position" (Žižek 2012, p. 359). Qualitative or perspectival dimensions supplement the mere fact of a sphere. Observing and making sense of the experience of others can further develop the qualitative or subjective experience of the sphere. For some there was a revelation in realising that the sphere could be understood or approached in many different ways and that the equivalence of these alternative approaches might be demonstrated. As their teacher, I opened many opportunities for ideas to be negotiated or exchanged. The students typically worked in small groups but regularly came together for whole class discussions. Every half an hour,

I typically stopped the group so that they could engage in 5 minutes of private reflective writing to capture as "live data" how their understandings were developing. On some occasions students were invited to read their comments aloud with view to revealing the diversity of response. For their "homework" after the session students were urged to build a commentary of the lesson out of these pieces of reflective writing combined with their written mathematical work. The story of "how the learning took place" for the individual and for the different subgroups was emphasised as a major lesson objective. That is, the pedagogical story was valued as well as the mathematical story in the steep learning curve entailed in the student's journey from seeing mathematics primarily as subject knowledge to seeing it as pedagogical content knowledge or as mathematics conceived from the point of view of a learner. The final package of material collected over 20 weeks allowed students to track their own learning history as evidenced in their evolving attempts to capture their mathematical experiences in reflective and increasingly critically analytical writing. In a separate study, I worked with a colleague in enabling graduate student teachers of English to document their own personal changes in relation to their subject area over a year as a key dimension of their course assessment (Hanley and Brown 2016).

On the occasion described, another student reported a more affective experience of being within the activity:

> On reflection, at that moment, I felt a real mix of emotions which combined many of the emotions that pupils face when asked to speak in class, in situations where they are not completely sure of what they are talking about. These involved almost a fear of saying the wrong thing, a desire to achieve the right answer, a wish not to appear foolish in front of the "teacher" and peers and also a concern over whether my explanation will be understood or even make sense.

This extract suggests that her mathematical experience was imbued with social, emotional and historical baggage. The imagination of the object in question, however, has real effect on the person's image of his or her self in attempting to make tangible the object that is being sought. The person is reflexively constituted through their attempt to grasp it. The grasp reveals aspects of who they are. Thus, the activity was centred on each student exploring alternative subjective positions, on documenting connections to alternative formations of self, such as a physical self moving in space, a pedagogical self reflecting on the learning of others, a geometric self creating drawings, an algebraic self who is solving formulae and an emotional self commenting on how it felt relating to other students in the context of the supposed mathematical entities. But in building these images of oneself, one is alerted to territory that one can grasp in a tangible way and seemingly to spaces beyond reach that can only be pointed to or speculated. There are questions as to from where things are being seen. What am I seeing? Who, when or where am I precisely to be seeing it in that way? What had been movements of the body became elements of one's comprehension of reality itself. The experience of the configurations became linked to how one felt at the time, a narrative of participation formalised for posterity, seemingly held in place by both rationality and belief at "the moment of pure subjective decision or choice which stabilizes a world" (Žižek

2012, p. 367) or implies a centre and specific sense of direction (Laclau 2007, pp. 66–83; Derrida 1978, pp. 278–294) or a new master discourse (Lacan 2007).

Each time these sessions are new for me in my role as the teacher as each individual account has unique features and a chosen storytelling style. Recent sessions have been enlivened by a move to the large brand new building that allows new spatial experiences – a huge internal atrium for suspending/stretching string in 3D, a drama studio that allows spotlighting, expansive flat space outside (where cold weather tempered the pleasure of making the curves very large), etc. My persistent ambition, the point of my teaching spelt out for my students, is for me to see things in new ways, to keep my teaching alive by enabling my students to provide new stimulus to each other and to me, to resist final versions that sum things up, towards recognising and perhaps specifying the limits of one's certainties and uncertainties. As Laclau (2007) argues:

> the abandonment of the aspiration to 'absolute' knowledge has exhilarating effects: on the one hand, human beings can recognize themselves as the true creators and no longer as the passive recipients of a predetermined structure; on the other hand, all social agents have to recognize their concrete finitude, nobody can aspire to be the true consciousness of the world. (p. 16).

In week 20 of 1 year's cohort, a challenge was set to carry out a photogenic activity for the university marketing department to show off our new education building. Matt Ballantyne, a student on the course describes his experience drawing on these photographic images (Figs. 6.4, 6.5, 6.6, and 6.7):

Fig. 6.4 Securing the end of the curve

Fig. 6.5 The author with a student admiring the suspended double curve

> Today we would be venturing out of the classroom and down to the Spanish Steps. Our task to create two curved lines that descended and intersected at one specific point using a large ball of string. We had to develop a collective vision of how to work on the task. After some thinking, loosening, tightening, adjusting and altering we were all happy with the placement of the string. The two strings formed smooth curves that intersected at a single point. Now the real challenge began. The mapping and representation of what we had created. There were many obstacles to overcome when creating such a 3D image. There were many different measurements that we needed to make. We considered the measurements from the front looking at the balconies and floors above. Another group went to the side and looked at it with the balconies to their left and the entrance to their right looking along the Spanish steps. The last group pretended to be birds looking down on the site and seeing the points on the floor. The group reassembled and shared their results. Having collaborated, we were able to see if any measurements that were needed to map the curve were missing. When we were all happy. We had what we needed to create an accurate representation of the curve, so we headed upstairs to create our model. We believed that not one model but many would best represent the situation. One group decided to draw three different 2D images to show the 3D planes which we had measured. The other two decided to try a 3D model. One group went for a sketch of the curves. Our last team were probably the most audacious and decided to make an actual 3D model including little people. They carefully crafted a scale model of the Spanish steps with string and cardboard (See Fig. 6.7). A single task, undertaken by a group who have worked closely together for six months, had taken completely different turns and ended up in many different places. The thing I like most about the end results is that they are ultimately a complement to each other. (Ballantyne and Brown 2016).

Matt's description reveals the way in which his apprehension of the curves evolved through a succession of physical movements, measurement activity, visual perspectives, inspection of data and shared interactions and productions with other students.

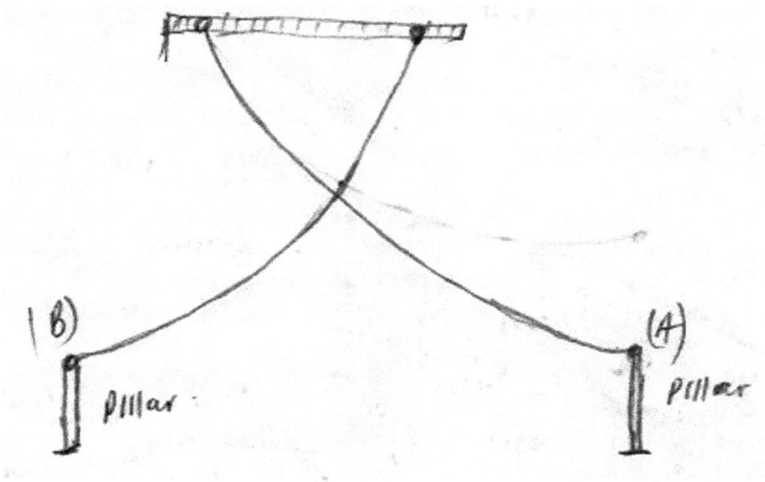

Fig. 6.6 Drawing of the suspended double curve

Fig. 6.7 Model of the suspended double curve

The description itself, later published in a teachers' journal, further processes Matt's comprehension of the mathematical entities concerned and stretches his capacity to articulate his understanding.

6.2 Apprehension of Planetary Movement[2]

Some of the bodily movement exercises conducted in our sessions involved acting out the relative configurations of earth, moon and sun and how these configurations might be seen from alternative perspectives, from deep space, from the surface of the earth, etc. (Heywood and Parker 2010). These configurations were enacted firstly with a globe for the earth, a small white sphere for the moon and a torch for the sun (Fig. 6.8). Later individuals took the place of the moon then spinning in relation to the earth (Fig. 6.9). The purpose of the sessions was to enable students to share their attempts to apprehend variously located mathematical objects, experienced as if navigating and orienting themselves inside big versions of the shapes concerned. That is, they told progressive stories of themselves, as apprehenders of the variously perceived spatial environments, developing technologies through which specific orientations could be achieved.

Some pieces of data derive from the research orientation of the sessions more generally. Everyone, students, tutors and visitors alike kept extensive records of their activity during each activity to better understand how the learning of mathematics happened. An extensive catalogue of video clips and photographs were collated, which were later used to build the written records that were produced. In the subsequent discussion, alternative approaches to framing school mathematical objects arose were suggested. From the teacher education perspective being taken, we sought to explore how alternative subjective positions can be productive of important qualitative aspects of the mathematical phenomena being portrayed. That is, these qualitative features, specific to the *world* in which the ideas were encountered, provided markers for observing and orienting the mathematical ideas being approached. School mathematics is typically encountered through qualitative features of the pedagogical worlds being entered (e.g. needing to make an argument to peers or tutors, representing ideas in different forms, reference to standard ways of depicting ideas for examination settings, etc.). Learning about the mode of embedding and working within it is part of the necessary learning required in many instances, especially those directed at supporting a specific purpose, e.g. conceptualising the moon that we can see as being on a circular (or is it elliptical) orbit.[3]

(i) The first piece of data comprises an extract from our written records collected during the activities as part of our own research:

> Kelly had brought some data with her, such as the exact duration of the day and the year, and it was apparent that her preparation for the activity was systematic, mentioning terms

[2]This section draws on material first published as: Brown, T., Heywood, D., Solomon, Y., & Zagorianakos, A. (2012). Experiencing the Space We share: rethinking objectivity and subjectivity. *Zentralblatt fur Didaktik der Mathematik: The International Journal on Mathematics Education*, *45*, 561–572.

[3]My students were far more graceful than the group of drunks doing a more complicated version of the exercise in Bela Tarr's movie *Werkmeister harmonies*: https://www.youtube.com/watch?v=_d5X2t_s9g8.

Fig. 6.8 The sun shines on a revolving world with the moon above

Fig. 6.9 Experiencing the movement of the earth in relation to the moon

such as "aphelion" ("which is the point of orbit furthest from the sun; ... which is going to be our winter"). As soon as the activity started Imogen said, 'the sun shines and the earth spins and when you don't have the sun on you it's night time'. Kelly pointed out that the length of a day is exactly 23.97 hours, reading it from the data that she had brought with her. ... Imogen replied immediately that 'there is noon when the sun is at its highest, when you are closest to the sun'. She gave an example by choosing Saudi Arabia on the globe and turning it: 'if we look to Saudi Arabia, it is noon in Saudi Arabia, as it moves away the sun is sinking again and then it goes to night time and then this is the midnight, and then it gets early, the sun is rising, the sun is rising, it gets to the noon, the sun is at its highest point'.

(ii) Another piece of data comprises extracts of reflections from an experienced mathematics teacher within the team researching how mathematical objects result from pedagogical exchanges as part of his doctoral studies. During the session, he was observing the students but occasionally found himself drawn into discussions as the students had known that he was quite good at mathematics, and in turn he could not stop himself from revealing his knowledge. In the reflections, the teacher was exploring the consequences of these unexpected interventions from a pedagogical point of view. The extracts refer to the sequence above. They were chosen with view to showing how the teacher's reflections are revelatory of his own identification with conceptions of pedagogy and of scientific discourse.

The following comments indicate his pedagogic orientation:

- I was kind of prepared for it.
- I don't want to "spoil the fun of discovery".
- I responded with an expression of approval.
- I pretended to agree.
- I instinctively tried to break the rhythm, so I said something that wouldn't be much of a clue.
- I repeated what Kelly said, trying to sustain Kelly's conclusion as a base for the subsequent investigation.
- Without realizing it I was entangled in the group discourse in the way that I was initially trying to avoid.
- I fully understand that [was] my old reflex as a teacher.
- I had fallen in the trap of influencing the group, as I could not disengage myself from its activity and as I interrupted the group's interaction to some extent; I became a victim of my own devices.
- These comments however point to a "correct way" of seeing things:
- She was not using "aphelion" *the correct way.*[4]
- Using "aphelion" and "perihelion" *the right way.*
- Trying *to keep the level of the group discourse as advanced as possible.*

(iii) In the final extract, Kelly, Imogen and Rebecca (Fig. 6.2) share their apprehensions of how the moon moves in relation to the earth. They experienced some difficulty in communicating these apprehensions in words. Finally, they

[4]The teacher reports that at one stage: "Kelly mentioned the summer time, introducing the term 'aphelion'".

enacted the orbit of the moon through bodily movements that seemed remarkably coordinated, with all three students moving in the same trajectory around a suspended sphere (the earth), where they each maintained a constant orientation to the earth throughout. Successive attempts interrupt each other:

K: Because we are on an angle of let's say this way I am looking at it… as we come round if we keep on that angle we only ever see my face, you never see the back of my head.
I: It doesn't matter whereabouts.
K: Yeah you split…
R: Kelly's focus stays on that ball so her body might be turning but she is still looking…
K: So you only ever see…
R: So, if someone is stood on there, they would only ever see my face, not the back of my head, otherwise I'd be going…
I: We must be right because we are all on the same wavelength. We all agree.
K: If I could spin myself like this …
I: The moon's just on an angle. That's what it is. Spin round double … see it's worked… best logic I've thought of.

The three examples comprise individuals displaying a range of pedagogical interests and attitudes towards notional mathematical objects. We have trainee teachers who oscillate between an unsteady grasp of the terminology and a more symbiotic immersion in the evolving world to which this terminology attempts to cling. This terminology is included in their own learning narratives within which meanings evolve. We have a teacher referencing his own interventions to established parameters. We have teacher educators in the background managing an activity towards influencing certain pedagogical results. We have researchers adopting more theoretical perspectives on how mathematical ideas are being framed. These alternative perspectives link to alternative conceptions of learning (discovery approaches, gravitation to correct understandings, creation of fresh perspectives, etc.) that variously construct and position mathematical objects and shape the apprehension of more or less familiar forms of knowledge. The enquiry in this chapter is specifically focused on how the participants variously identify with specific conceptions and how those identifications support teacher education ambitions, specifically those relating to building narrative around learning experiences. We cannot assume any sort of correct overview of the activity that took place nor be representative of the multitude of insider perspectives.

In the first extract, Kelly's experimental introduction of specific terminology is depicted as the consequence of advance preparation at home, preceding a more settled understanding of the parameters that framed the terms that she used. "Aphelion", as an embodiment of, or subjective perspective on, an ellipse, for example, was occasionally asserted as being linked to a position on an orbit closest to the sun, rather than furthest. Yet the bigger point is that the *world* that would host this term within a more secure scientific discourse was apparently not yet in place for her. Neither the host space nor the objects it allowed had been established. The technical term "aphelion" for Kelly was dislocated, floating in space as it were – its home had not yet been fully conceptualised as a point within an overarching spatial structure. Yet clearly, she was introducing the term to provisionally mark out the territory that she was seeking to better understand.

In the second piece, we view the events through brief extracts from the teacher's extensive reflective writing where he indicates his own unexpected participation. The extracts point to a specific mediation and more or less obliquely depict his involvement in the activity. The teacher's supposition of the task in hand is at least partially centred in a particular conception of the knowledge to be apprehended. Yet this interest is obscured by his own concern that he be an observer rather than a teacher. This is against the backdrop of Imogen, Kelly and Rebecca playing out alternative approaches to the task where they have prepared for the task differently and get to be convinced differently when they come together on the task. The routes from their understandings to their rationalities are different in each case. The teacher has an ambivalent role focusing primarily on understanding how the others are apprehending the task, where his own involvement in the proceedings is nevertheless having some impact. The ideas in question are manifested differently through the thoughts, action and speech of the people in question, in relation to a set of activities designed with certain pedagogical ambitions in mind. But the issue for this chapter is not with the relative merits of the perspectives achieved but with how mathematical objects derive from alternative subjective positions or modes of identification.

In the final example, the mathematical object in question is a circle (or ellipse) but where many qualitative dimensions of the pedagogical world supplement the students' experience. The sensual perspectives assumed of this circle obscure its appearance as a clear cut geometric entity. The task was centred on being able to apprehend an orbit from various given perspectives, such that the students were challenged to situate themselves *within* and experience mathematically conceived space. The question of moving around this ellipse whilst maintaining the correct orientation further complicates the sharing of perceptions in words. The keenly felt perception of being on the "same wavelength" within shared movement, however, somehow reduced the need for a clear set of words – the agreement could be danced. Indeed, the desire for a correct set of words seemed destined to fail as the power of shared movement became far more evocative of the entity in question. (See also Roth and Thom 2008). The students are identifying with an experience that defies final capture in a symbolic form, but it also defies final capture of the students themselves in finished form. Their subjectivity is referenced to a lived experience, with no fixed relation of object (an elliptical orbit sought through a succession of fragmented sentences) to subjects (held by names and relations to other subjects). In a "real-life" context the affectivity of the space teaches the students to recognise their position in time and space through sensual clues, (e.g. shadows, direction of moon, darkness, temperature, reciprocities of sharing space with others). Their emergent spatial and temporal awareness, marked by these qualitative features, occurs as part of a layering of complex systems of relationships and spaces within constantly changing circumstances.

6.3 Lockdown Mathematics

This section of the book is the last piece to be written (in June 2020). Eight weeks into the 20-week course, England went into lockdown. We did not know that the eighth session would be the last and the course reverted to sharing hastily composed emails. After several false starts, we managed to get Zoom sessions working so long as my wife did not have her Zoom Pilates class at the same time. The new format became that I set a challenge first thing on the morning of the session and asked members of the group to do a one page write of their findings and to send this to everyone in the group. Figure 6.10 provides a brief extract from one of the pieces of work circulated by one of the students at this stage of the session. The variety was considerable among the ten students as other examples will show. When we congregated on Zoom at 11 am, each person talked about what they had done for a few minutes. This generated many ideas and questions. A key dimension of the discussion was to consider the investigation as a mathematical field of enquiry. The metaphor I use is for students to imagine that they are the finance minister facing a very complex situation but where it is known that various strategies work but in a conditional way. For example, we know that at the micro level a factory owner can reduce her production costs by paying the workers less so long as she can hold on to her workers. Trouble is that if every factory owner does that at a macro level, then workers cannot afford to buy the products, and the factory owner makes no profit (Harvey 2015, pp. 112–130). Consequently, the finance minister or factory owner needs to play several ideas off against each other to improve their respective successes. This contrasts with the approach to a school mathematical investigation described by Tanuj Shah at a link provided below where he solved the rectangle diagonal investigation outright, something that most school children could not do. Nevertheless, I have found that the activity is quite feasible with a mixed ability class of 11-year-olds, where there are many dimensions to the activity that allow all children to participate with various more qualitative aspects of the challenge and where sharing alternative perspectives is the chief pedagogical benefit. It was this attitude that I was seeking to explore with my adult students. The challenge was for each student to find out the way the task was understood by others. In presenting their thoughts, students need to articulate their understanding and in so doing strengthening and developing their understanding whilst enabling their partner to have access to these thoughts. But

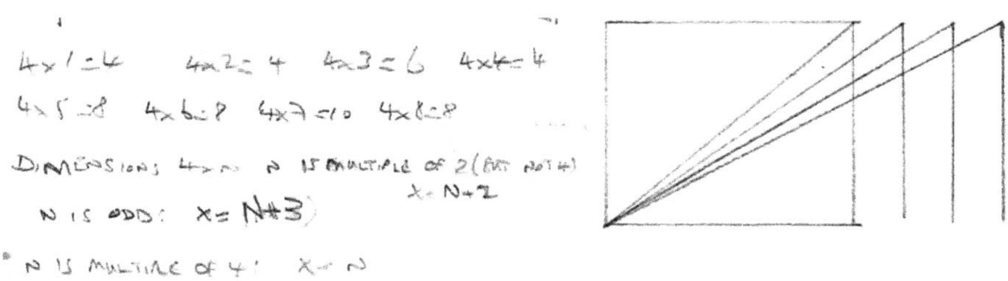

Fig. 6.10 Extract from notes shared by Khurram before Zoom meeting

Fig. 6.11 Dean doing a homework on curves. (Short film available on Youtube: https://youtu.be/6cN3oRcekPA)

mathematically, the task never really reaches an end if this approach is taken. It is not about reaching a final solution but rather exploring the what ifs of the situation as a way of building more sophisticated mathematical dialogue and independent thinking. The territory may never be fully conquered, but one's navigational skills in the field have been greatly enhanced. In the session, we talked and worked things out mentally until we felt our brains had reached capacity. The attempt at 3D cuboids was the final straw, and it was suggested that we retreat to our private space again so that we could write things down. I include below the text of an email that I sent to my students shortly after where I was explaining what I was attempting to do and summarising some of the points I thought we had discussed in the session. Last term I encouraged a write from the students every week. This term I am asking for four extended pieces of work of their choice over the 10 weeks of the second half of the course. I am anticipating that some of the students will choose to do an extended piece on this field of enquiry, where they will combine mathematical thinking, presentational quality, discussion of their own learning, how this contrasted with the work of others, etc. (Figure 6.11 provides a still from a film submitted by one student for his assessment at the end of the course).

> Last night I did a Google on "rectangle diagonal investigation" and got this:
>
> https://apfstatic.s3.ap-south-1.amazonaws.com/s3fs-public/07-tanuj_diagonal_investigation.pdf
>
> I did a screen shot of the first page and sent it to you. I have not done it for a few years and could not remember how productive it was as an activity, but I was hoping that I would be surprised.
>
> I was. I received a good range of work from you where the activity had been understood in a lot of different ways but with some core similarities.

Then the discussion made us all dig a bit deeper, and we came up with a range of questions that potentially defined multiple directions in which the activity could be further developed. This is PhD researcher in mathematics type mode, not the typical school mode where so often the teacher asks a well-defined question and gets a right or wrong answer back. In this activity there was no scope for being wrong, just seeing things in different ways, but with a firm footing in things we knew for sure. The algebra emerged as pathways become more familiar – as generalisations could be seen.

I was really taken by James' introduction of Pythagoras – something I had never thought of before. At first I thought James was a bit off track, but after a while I realised that James was seeing it in a more sophisticated way than I was. In short, for very large rectangles, the length of the hypotenuse gives you an approximation to the number of squares passed through. That provoked my question – how do you fix the dimensions so that you maximise the number of squares passed through?

Ellie had this great line: "different rules for each prime w and H value therefore infinite primes = infinite rules". Mind blowing, but is that true??

Ben's progression through successive heights for rectangles seemed to suggest that there was a pattern to how many squares were passed through in successive rows. Could that lead to a formula where you could find the maximum number of squares that could be passed through? We know the minimum number is equal to the longest side. But what about the maximum?

So we understand simple examples very well – but we are more foggy as cases become more complex – but we have resources to tackle those mysteries.

If you choose to write this up you could:

- Explore the limits of your own understanding now.
- Consider the differences between the approaches taken by different people.
- Pursue a line of mathematical investigation from things we have found out so far.
- Ask yourself the question – how is this activity stretching and developing our mathematical capabilities?
- If you were faced with a mixed ability class of 30 11-year-olds, how might you organise the lesson so that everyone is included for the whole lesson?

Chapter 7
Rethinking Objectivity and Subjectivity

7.1 Introduction

This chapter develops more theoretically the key assertion of this book that our conceptions of mathematical objects are functions of how we conceptualise the human subjects apprehending them. Mathematical objects, such as those presented in schools, are positioned within ever-changing forms of life that constantly reposition or reconstruct those objects and the meanings conferred on them. Meanwhile, in the model proposed, human subjects are understood not primarily as biological or psychological entities but rather in relation to evolving or shifting discursive backdrops that can change *who they are*. The chapter then is centred on an interest in understanding how life, and specifically pedagogical activity, produce and confer meanings on mathematical objects and on human subjects, rather than supposing that those objects and subjects precede the turbulence of life as preformed entities. We need to see mathematical thinking as more than a field of knowledge where experts are assumed to be finding out about the gaps in that knowledge. New knowledges disrupt the functioning and territories of old knowledges. The life that we lead prevents knowledge from being stable except in localised ways. Any assertion of such localities restricts our capacity to conceptualise and occupy new ways of being in worlds hitherto unthought.

Our positioning as teachers and students confronting mathematics responds to newly emerging manifestations of mathematics. For example, machines occupy spaces previously held by human operatives (e.g. cash tills totalling purchases, programmed automatic market trading, robotive factory procedures, medical technology, computer-centred mathematics and calculators). The materiality of the human reach needs to be understood as incorporating such apparatus (Barad 2007). Palmer

---

This chapter draws on material first published as Brown, T., Heywood, D., Solomon, Y., & Zagorianakos, A. (2013). Experiencing the space we share, *ZDM: The International Journal of Mathematics Education*, *45*, 561–572.

T. Brown, *A Contemporary Theory of Mathematics Education Research*,
https://doi.org/10.1007/978-3-030-55100-1_7

(2011) has shown how Barad's notion of apparatus underlies the very structure and apparatus of schooling that supports mathematical learning (e.g. school building formats, squared exercise books, registration requirements, pedagogical models, curriculum frameworks). Hoyles et al. (2010) have shown how humans in the workplace now need to think systemically and not so much engage in detailed mathematical operations. Meanwhile, as seen in earlier chapters, initiatives such as curriculum implementation in education and associated assessment impact on how a given community builds its wider public understanding of mathematics and of associated technology/apparatus in ever-changing circumstances. Those pedagogical practices ultimately come to define that community's conceptions of mathematics and how that community expresses its demands on educational processes and hence on teachers, in those areas.

This chapter draws on some contemporary philosophy, especially Alain Badiou, where conceptions of object and subject are brought into a new relation, as I have previously discussed (Brown 2011). Badiou (2007, 2009, 2011) is one of France's leading philosophers. He rejects erstwhile distinctions between analytic and continental philosophies through embracing both the technologies of the former and the more temperamental conceptions of subjectivity associated with the latter. On the one hand, Badiou builds a new conception of "object" that results from fitting new models to newly supposed worlds. He alerts us to the contingency of hitherto supposed worlds and the objects that they support. Meanwhile, Badiou invokes Lacan's psychoanalytic theory. Here subjectivity is predicated on a more collective conception of the subject, where an individual is understood with respect to his or her collective participation in the name of some wider adjustment. Specifically, as an example to be explored here, one can adopt a range of attitudes or identifications to supposed mathematical correctness in pedagogical situations. Such situations are built around suppositions as to how mathematics provides an analytical frame through which to contemplate our lives and around alternative pedagogical assumptions as to how ideas are constructed and shared. These different modes of subjective identification display alternative pedagogical attitudes but also result in mathematical objects being produced differently in notionally shared situations. There is a challenge to understand how emergent mathematical thinking can be activated and approached through pedagogical interests. Specifically, teachers will not be adequately prepared for future teaching with past versions of knowledge. They need to be responsive to new ways of thinking that will locate mathematics in new relations with life. We shall specifically counter the idea that a teacher needs to understand new challenges in advance of her students.

A characteristic feature of much research in mathematics teacher education is that it is conducted within one of the three relatively distinct fields of teachers' knowledge,[1] teachers' beliefs (or affect more generally)[2] and teacher identity.[3]

[1] For example, Hill, Rowan and Ball; Rowland and Ruthven (2011).

[2] For example, Zan et al. (2006); Hannula (2012); Walshaw and Brown (2012); Roth and Walshaw (2019).

[3] For example, Black et al. (2009); Walls (2009); Walshaw (2010).

In some more recent instances of research, affect is understood in terms of how the trainees experience the demands to participate in emergent professional patterns of discursive activity.[4] The meta-discussion proposed here relates to these recent approaches to affect by embracing each of the three fields in relation to Badiou's model. The discussion reconfigures knowledge as compliance with specific models of mathematics and of mathematical learning. Affect is understood in terms of resonance or dissonance between the individual's sense of self and the model to which that individual feels obliged to conform. Identity is recast as successful or unsuccessful *identifications* with discursive formulations. The cognition/affect interface (McLeod 1992) is displaced, in crude terms, by subjectivity being referenced to identification with such narrative accounts shifting through time, rather than on the functioning of individual brains in a given situation. The meta-discussion links the trainee teachers' mathematical experimentation to their participation in a permanent state of adjusting to new conditions, where neither brains nor mathematics precede life. There may be affective consequences, or plain awkwardness, in adjusting to new forms of knowledge. Yet such is life. The awkwardness is not something to be abolished. Similarly, in reading Hegel: "There is no reason that we should indulge the desire to heal when we read Hegel" (Pahl 2011, p. 14). Rather, new conceptions of mathematical knowledge, such as pedagogic framings introduced through new curriculum initiatives, or schematic approaches popularised through work or leisure activities, feed into a collective working through of these conceptions, which make qualitative adjustments to that mathematical knowledge. It is for the new generation of teachers to work out what those new conceptions mean for them personally as they negotiate their path into teaching and subsequently how they might tap those new conceptions as pedagogical opportunities with their future students.

In the previous chapter I drew a comparison of mathematical knowledge and knowledge in economics. The (classical) economics that I taught myself in high school seemed to be very different to the (Keynesian) economics I did at university, which was again very different to the economics my son did in high school. On TV this week, I watched a discussion between the (right wing) Scottish historian Niall Ferguson and Stephanie Kelton, a (left wing) American economist, and they simply could not agree on how to solve a very specific economical problem concerning an attitude to building up a deficit. Mathematical knowledge does not change or differ between people in quite the same way. There are changes in mathematical knowledge, but those changes are more down to changes in subjective perspectives on, or selections of, the mathematical content, not so much with the mathematical content per se. I have mentioned how exam formats change on a regular basis through history, each successive versions framing school mathematics in a different way according to the perceived priorities of the day. But the qualitative differences in the perception of mathematical content are very important within pedagogical processes, rather than supposing that the final right answer is the whole point. Of course the final right answer has a certain poignancy if you are sending Elon Musk to a space station, but for those of us working with students in schools and colleges, it is

[4] For example, Brown and McNamara (2011); Roesken et al. (2011); Walshaw and Brown (2012).

important to attend to the qualitative environment through which they learn capabilities that ultimately result in them producing right answers. Put simply you don't know what a right answer is until you have seen a wrong answer or two. But asserting the right/wrong dichotomy or, to be more technical, distinguishing between conceptions or misconceptions is not an especially motivating approach for students to learn how mathematics works or to get a feel for it. To process questions as having clearly right or wrong answers is to exclude from discussion crucial qualitative understandings of both mathematics and people.

The chapter commences with a brief outline of the university classroom situation in which these themes are explored. A sketch of some theory is then provided as a prelude to two sections, which in turn introduce conceptions of subjectivity and objectivity derived from Badiou's work and referenced to the classroom activity. I then provide some examples of research data centred on the negotiation of mathematical objects. This data is discussed from the point of view of how the depicted participants are variously positioned in relation to the mathematical models in question. This provokes a question as to how new teachers might conceptualise the objects that they will eventually teach, as objects to reproduce or as objects to renew.

7.2 Setting and Aims

A key theme of this present chapter concerns how participants variously identify with conceptions of mathematics and how those identifications support teacher education ambitions. As discussed in the previous chapter, the author has collected data over many years with successive groups of postgraduate students training to be mathematics teachers in British secondary schools. The product of this activity has provided data on how students conceptualise mathematical ideas. In each of the years, I have been teaching 20 3-hour sessions to each group. As discussed earlier, the sessions were each designed to broaden the student teachers' conceptions of mathematics through carrying out a variety of mathematical investigations, to see mathematics from a broad range of unfamiliar perspectives. Through such activity, the students were encouraged to explore themes independently, pose and answer their own questions and reach mathematical generalisations where possible. The agenda of the sessions is set out as being centred on all participants (students and staff) researching together how people build mathematical understanding. We wanted to know how people learn mathematics. What might our shared learning (as mathematicians, as pedagogues, as researchers) tell us about this? How might the students' analytical approaches be developed and transferred to their work in schools? This enables the production of data (such as reflective writing produced during and after the sessions, multiple sound and video recordings, alternative approaches to the mathematical work, etc.). The students were encouraged to submit a file outlining their research for their end-of-year assessment. Two or three of the sessions each year have been devoted to the apprehension of geometric entities through exercises centred on the students' own bodily movement (Brown, T. 2011).[5] The exercises became a prelude to the students formulating mathematical models of the configurations they had encountered in these physical exercises as part of thinking through how mathematical entities come into being for themselves and potentially for other students.

7.3 Identification

I now turn to a consideration of how identification might be understood. Humans progressively work between the physical world that they apprehend in everyday life and conceptions of that world derived from more socialised ways of making sense of that world. Here, I am especially interested in those ways pertaining to more mathematical accounts of the world, as defined through the symbolic apparatus typically utilised in such accounts, and in turn with how mathematical objects are used to support those accounts. Elsewhere, I have suggested that such mathematical accounts presuppose ways of looking and in this sense shape the parameters of what it is to be a human subject (Brown, T. 2011). I developed this idea in Lacanian terms where individuals have a common-sense view of the world and of themselves within it, through which they apprehend objects and their own relationships to them. That is, individuals initially understand the world, and themselves, through this common-sense view. We considered the case of the mastery curriculum earlier. Yet, acceptance in the shared world requires a negotiation of the symbolic networks (such as pedagogical apparatus) that have been produced, by those who have preceded us to make sense of the physical world. The scientifically defined universe contingently defines worlds and the human's place within them (Lacan 2008). It may, however, be that the individual is not especially comfortable with these assigned places and that there are consequences to these perceived failures of fit. For example, psychology defines individuals in terms of various physical or responsive attributes, which may bypass the affective sense of self possessed by the individual herself. Or alternatively, the individual human might too compliantly accept this external designation. Lacan's model locates life as a negotiation in which the individual works through successive accounts of the world, each of which points to a place for the individual. He mocks the failure of scientific constructs to keep up to date, consigned as they are to the need for regular renewal, whilst in his view the human always survives. It is rather like students never quite getting to the end of the rectangle diagonal investigation from the last chapter but can only ever iteratively improve the previous model.

What does this look like in the specific example that we plan to address here? Humans experience geometric objects in orienting wider spatial awareness, and that empirical site enables individuals to produce or share mathematical objects.

[5] On the occasion being described, the team comprised the two regular teachers, a science teacher educator initiating the specific activity and an experienced teacher conducting PhD research, and a video operator.

Empirical reality here, however, is understood as being produced through interpretive procedures derived from specific understandings of human subjects and how they frame their sensual experience. That is, empirical reality is just one version of events that fixes life in a certain way. For example, humans start, inevitably, with naïve ways of apprehending the moon, the sun and the stars. They progress through a more intuitive sense of how things work – *the moon moves during the night.* Then perhaps they encounter a mathematical frame of reference for a more shared human knowledge – *the moon encircles the earth but we only see part of that circular move.* This shared human knowledge takes different forms in different educational locations, according to priorities, level and so forth. Mathematical knowledge, for example, depends on research funds motivated by current agenda and more immediately in schools on decisions to include selected aspects in the curriculum as processed through different pedagogical modes.

7.4 Subjectivity

Lacan's (2006) approach emphasises the societal demands that shape the individual human subject. The subject derives from the stories that are told about him or her, or from the stories that are told about people, or classes of people more generally. The individual may or may not like the way in which they are being classified. In the framework that we are following, the mathematics teacher's identity is a function of how the teacher is understood in a given location or time, perhaps according to the skills, competencies and practices seen as "normal". Learning here might be understood more as being about an experience through time rather than being about apprehending mathematical ideas located in a fixed conception of space. Education comprises the formation of objects/events in time/space rather than being about an encounter with ready-made entities. Mathematical ideas cannot necessarily be apprehended in an instant. They may have a time dimension, as a conceptual process (Tessier 2012), or through their location in an unfolding historical development (Corfield 2012). The apprehension of an idea may result from a gradual assimilation of the idea's components and qualities and how these are combined in its formation. I may compare new sets with a selection of previously known sets. I may contrast the operation of a newly located function with more familiar functions. The progressive apprehension of the supposed idea becomes part of the story of my life, a part of getting to understand who I am and how I fit into a supposed world or how I might make that world otherwise. That is, this progressive apprehension builds a story around the abstract entities being located, a nuanced qualitative layer in which any learner is fully implicated since it was integral to their very own constitution – a supplement in the story of their lives. The individual's actions comprise part of a collective response to such situations. This collective response might result from mathematics being viewed differently more generally, for example, consequential to curriculum change, through mathematics being seen differently in popular mythology, in changes to the demands on mathematical capabilities and so forth.

Badiou draws on Lacan's conception of subjectivity. The subject, rather than being seen primarily as a biologically framed cognitive entity, is understood through a reflection of a broader symbolic universe. Lacan's concept of human formation is triggered by a transformation that takes place when a young child assumes a *discrete* image of herself. Lacan's iconic example is that she looks into the mirror and recognises herself. This allows her to postulate a series of equivalences, samenesses and identities, between herself and the objects of the surrounding world (the equivalence of my movement on the floor, to the drawing on paper, to the image in my mind, seen as continuous movement, or as a configuration of points).

For example, student teacher Imogen, first met in the last chapter, carried out a body movement exercise in which she tried to maintain equidistance from her body to a fixed point and to a straight wall. She commenced by being positioned halfway between one of her friends and a nearby wall. The first part of the activity comprised attempts to physically move from one point to another maintaining the equidistance. This challenge was shared with three peers all of whom had different views on how Imogen might achieve this (or not). After much discussion and walking around, a set of points was marked out on the ground using screwed up pieces of paper. The whole episode was videoed for later analysis. A rough drawing was created in which the points were joined. At home Imogen extended her notes. After further rough drawings and calculations, she eventually drew a graph featuring a point and the wall on to squared paper (with the fixed point being the origin and the line $y = 10$ as the wall) and used Pythagoras' rule to generate positions that met the criteria. The second point was located by drawing a triangle (0, 0), (4, 0) and (x, 4).

> From this we can pull out a triangle in the hope it will help us calculate what the x coordinate would have to be in order for the distance from the wall to the origin to remain equal. Using Pythagoras theorem we know that:
>
> $a^2=b^2+c^2$,
> $6^2=4^2+x^2$
> $x^2=6^2-4^2$,
> $x^2=36-16$,
> $x^2=20$,
> $x=\sqrt{20}$.
>
> We can now see that the coordinate for the new point should be at ($\sqrt{20}$, 4) for the distance from this point to the wall and this point to the origin to be six. We also know that the points will be symmetrical in the y-axis. So now another point will be ($-\sqrt{20}$, 4)

Imogen then plotted those two points on a graph. This was followed by her finding the x ordinate for the points at a distance of three, generating the points ($\sqrt{40}$, 3) and ($-\sqrt{40}$, 3).

> Carrying on this method, I continued altering the y coordinates so that the distance from the wall changed which in turn changed the length of the hypotenuse and also the height of the triangle; this gave us many more different coordinates where the moveable point

> could be so that it was equidistant between the fixed point and the wall. However, I did think that once the moveable point passed the x-axis, then there wouldn't be a point that would be of equal distant to the wall and the origin; however, after drawing a diagram and extending the graph a little bit more, I came to realize that it was possible for it to be below the fixed point as it was just that, a point; however, we could not have a point above the wall as the wall continued on for eternity so the moveable point would always be closer to the wall.

The calculations were combined into a table. As she developed more summative results over time, writing of this sort provided a narrative spanning 12 pages of notes, calculations and diagrams that documented her shifting perspectives from enacting physical movements on the floor to creating more formal diagrams and equations. This work thus provided a narrative of the student teacher's journey of learning during which the curve came into being for her. In the perspective that we are pursuing, such narratives document human subjects and mathematical objects coming into being. By creating such narratives in this and other sessions, the student teachers become more adept at accounting for their own learning process, making sense of who they are and how they fit in. The narratives on the process of emergent understanding provided excellent material for discussing and comparing learning experiences in our group sessions. The discussions enabled more refined use of mathematical terms, but more importantly the discussions provided a forum for considering more generic pedagogical terminology, such as "generalisation", "conjecture", "logical sequence" and "proof". Consequently, the student teachers became better able to report on the learning of their own students in a more refined language when they tried out similar activities in schools.

More theoretically, according to Lacan (2006), an image of oneself fixes an egocentric image of the world shaped around that image of self – like a personally constructed master discourse of who I am. That is, the assumption of a oneself results in a supposed relation to a world and a partial fixing of the entities one perceives to be within the world. The self is understood through being gauged against this supposed world. Initially, in our case, Imogen builds a sense of such relations by moving herself around the physical space. Imogen's sense of herself is referenced to instructions that have guided her movement. In due course these relations become implicated in more overtly mathematical phenomena that underpin her more formal approach. This shift of perspective comprises reflective awareness of symbolised relationships, such as how specific bodily positioning responds to a coded spatial environment. These objects are linked to "mathematical knowledge" and become relatively fixed with consequential restrictions on how relations between people and geometry can be understood. Imogen's assumption of herself comprises a collation of a set of characteristics, attributes, organs, positions, etc. that make up that self. This set of characteristics is "counted as one" person. Lacan, however, cautions that we should be wary of this image, since it is illusory. It is a snap shot that never quite works. It never fully captures the real me as it were, rather like the production of a rational formula not fully capturing the experience of understanding derived from moving according to the locus of a curve. In Lacan's model the limits of our "real" self are never fully visible to us.

7.5 Objectivity: Counting as One

Badiou follows philosophers such as Bachelard, Lakatos and Althusser in seeing science as a *practice* marked by the production of *new* objects of knowledge (Feltham 2008, pp. 20–21), in much the same manner as Deleuze and Guattari see philosophy as "the art of forming, inventing, and fabricating concepts" (Deleuze and Guattari 1996, p. 2). Badiou commences his analysis with a sheer multiplicity of elements in a pure state of being. His set theoretic approach locates a mode of organisation with no empirical reference. Here, there is no overarching unity, such as the oneness sometimes celebrated in theology. In this state, the elements are not anywhere but can be combined in subsets of that multiplicity to create or define unities. Badiou's assertion is that any such unity, or object, derives from an *operation* of "counting as one". That is, an *object* is produced by the operation of counting a set of elements, within a *supposed* world, as one object. These elements could be atoms, blood cells, GPS coordinates, emotions, humans or items on a mathematics curriculum. This operation brings the object (kettle, mouse, Swaziland, schizophrenia, the Manchester United football team, mathematics curriculum) into existence within a *world* (kitchen implements, rodents, Africa, health conditions, the English Premier League, schooling). And, in a sense, it also brings the world into being. The assertion of an object asserts the world that is the outside of that object, a world that has perhaps been changed a little by the specific noticing of the object. The *world* is itself a result of a wider "counting as one" (of the total elements of that *world*). In this formulation, any element can itself be a set and a potential member of other sets. And within any assertion of a set, yet further possibilities are created, resulting from the construction of subsets or power sets producing yet more new entities. This very proliferation itself defies any final stability in the universe. For this reason, there can be no settling or convergence in the meaning of the constituent terms. Badiou contemplates a partially managed multidimensional infinity. Yet forms of knowledge are predicated on a world, comprising specific sets of terms within this world. Such forms of knowledge might be disrupted as they readjust around the ever-expanding set of sets being counted as one. The advance of mathematics can be seen as the practice of producing its objects. For example, Badiou cites the introduction of i as a disruption to the conception of number. Such expansion reveals objects not previously identified within earlier overarching multiplicities.

How might this approach support the exploration of learning or more generally the human apprehension of mathematical objects? Mathematical thinking can generally be understood through the pursuit of noticing or asserting a generality, a notion resonant with "counting as one". The construction of a model results from an *operation* that apprehends, or perhaps creates, a set of elements as a unity. To continue our example from the last section, after nine pages of calculation, further maps and reflective writing, Imogen convinced herself that she could carry on producing points. Finally, she plotted the points and joined them to produce a curve. That is, the points (a–i in Fig. 7.1) were *counted as one* set, which Imogen finally concluded marked out the course of a parabola with an equation of the form $y = 5 - x^2$.

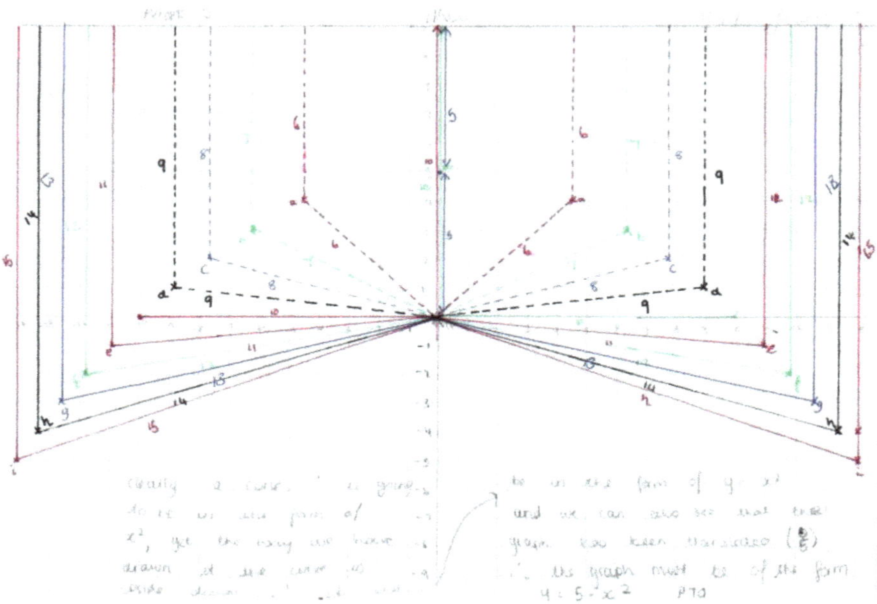

Fig. 7.1 Imogen's construction of a parabola

As she puts it, rather speculatively below the diagram: "clearly a curve, ∴ is going to be in the form of x^2, yet the way we have drawn it the curve is upside down, ∴ it will be in the form of $y = -x^2$ and we can also see that the graph has been translated, ∴ the graph must be of the form $y = 5 - x^2$". Her remaining pages of notes include an attempt at giving the general formula $y = a/2 - x^2/2a$.

In another example a student has not quite achieved the same resolution in translating the experience of walking a path into a drawn depiction of that walk. The diverse outcomes would normally be shared and discussed towards the end of the session so that the students could resolve differences in their write-ups at home. In his write-up at home, this particular student recognised that his account was different to some of his peers but had picked up that the curve was called a parabola but not yet linked that name to its process of construction (Fig. 7.2).

Badiou's notion of "counting as one" works whether we are considering students encountering socially known ideas for the first time, such as a parabola, or new innovations by researchers. A "counting as one" seen as the acquisition of a new model could be understood in either of these two situations in relation to a newly extended situation. The assertion of a given entity entails an operation to "count as one" the objects of a given set. But thereafter the term can become a member of other sets of objects such as "conics", e.g. parabolas, ellipses, circles, etc., seen as making up a world and utilised in organising our apprehension of the world. Algebraicisation comprises a similar operation of "counting as one" (e.g. identifying the set of points obeying the relation $y = 5 – x^2$). The objects get to be there, in a world, consequential to the operation. But they need that prior (or simultaneous) construction, of a world (in this instance two-dimensional space,

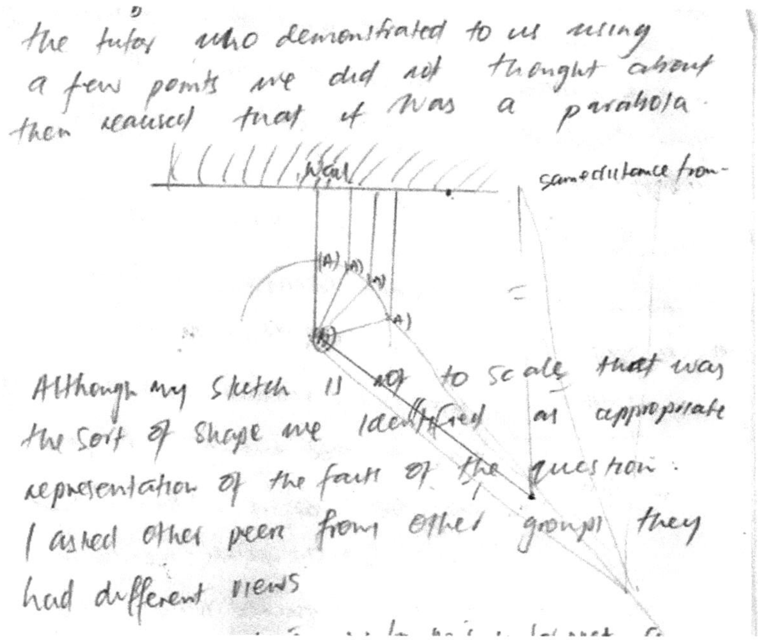

Fig. 7.2 Another student's construction of the path that had been walked

structured according to some rules), to *be there.* The existence of an object requires a place for it to exist. Badiou distinguishes between mathematics as a domain of truth and mathematical knowledge pertaining to a specific conception of a world. For example, geometry is knowledge if it is predicated on a Euclidean, or human, construction of the physical space in which we reside. Truth is eternal (but not static), whilst knowledge is forever being updated (as in the rectangle diagonal investigation) to newly perceived conditions (when algebra or correct answers become so very important) but at any point in time frames our perception of the world, as we know it.

Learning comprises the formulation and positioning of an object in a *world.* This requires the assertion of an object and an assertion of a (transcendental) world. Object and world imply each other. For the students moving around according to geometric loci, the task is to apprehend continuous movement as a sequence of points. These points are then aggregated to "count as one" object, understood in terms of this mode of aggregation. Retroactively the students can recognise the shape they have walked against a new register and see it as an object.

In short, an individual human (a set of attributes counted as one) confronts an object (made of elements that have been counted as one). These two entities come into relation in the given supposed world, for now. Yet the operation of "count as one" can always be performed differently according to new circumstances. The relation is contingent on a world that is always changing and needs to move on. This "moving on" underlies the concept of pedagogy as participation in the adjustments to life being addressed in this chapter.

To recap, a central tenet of Badiou's work is that an object must *be* in a world for it to *exist*. I have discussed in more detail Badiou's set theoretic distinction between *being* and *existence* in relation to school mathematics (Brown, T, 2011). In this approach, the objects of school mathematics are functions of implied worlds, whether those worlds are "real-life situations" or "mathematical domains" with their specific modes of functioning and inclusions. In this sense, *all* school mathematics is embodied. School mathematics is a function of institutional contexts and regulated as such. Barad (2007) has shown how scientific phenomena more generally are functions of the inspection apparatus through which they are viewed. Shulman (1986) famously made a distinction between *subject knowledge* and *pedagogical content knowledge*, whilst other writers questioned whether this distinction was valid since all subject knowledge is itself a form of representation (e.g. McNamara, 1991). A key argument of this chapter is that pedagogical contexts (SK and PCK) define their objects. Indeed Badiou (2007, p. 7) takes the extreme view that "*there are no* mathematical objects. Strictly speaking mathematics *presents nothing*". This surely applies to Badiou's use of set theory. The growth of geometry however has been shaped around empirically motivated objects, such as a circle. It is not easy to sort mathematics entities according to whether they are empirically referenced or not in their historical formation, and thus had narratives built around them.

7.6 The Ontology of Mathematical Objects

The set of people present are each assessing the domain according to their own respective perceptual capacities and according to the demands being made. They are each apprehending objects in potentially different ways, more or less, from a pedagogic perspective. But to what extent is it meaningful to speak of them as sharing mathematical entities in some absolute sense? There is an experience through time within the episode depicted that is unique for each individual, yet clearly there is some orientation around supposed points of sharing. In our example, Imogen developed her conception of a parabola, without ever naming it as such, through discussion and shared activity with some peers. But how might we understand such sharing? The philosopher of hermeneutics, Paul Ricoeur (e.g. 1984), argues that the passage of time does not lend itself to being described as a sequence of events, features or stages but instead needs to be understood as being mediated by narrative accounts of such transitions, relying on interpretations, which at a very basic level cannot be seen as comprising phenomenological features. The perceptual or phenomenological mark-up is different for each person at each point in time. Each has a story to tell. The mark-up is a function of the individual's specific identification with the wider discourse. A book edited by Doxiadis and Mazur (2012) brings together a set of papers each concerned with how mathematical experience might be understood through narrative where a time dimension to mathematical conceptualisation is highlighted.

How might we resolve the ontology of mathematical ideas in a school context? Is it possible to think of school mathematical objects as being outside of some sort of agenda? For example, the conceptions of students in England doing advanced level examinations at 18+ are rather constrained by the way in which questions typically frame conics, such as a parabola. In school mathematics, geometrical entities are normally presented as if from an objective perspective within a limited set of frames. The idea of subjective perspective, seeing a shape as if from being inside of it and moving around in it, as in the exercises with my students, would be rather peculiar in this setting. Perspectives are regulated. The mathematical entities are required to assume specific modes of existence for assessment purposes. As emphasised, Badiou's approach to ontology resists notions of primordial unity in favour of multiplicity, comprising elements in a pure state of *being*. This multiplicity precedes any notion of primordial relations, or objects. To *exist* these elements must be conceptualised within a "world" in which relationships between elements can be understood and objects can exist. The name "parabola", if it is known, can be assigned to a walked path or to a pencil line on a sheet of paper. Each world has a logic, but our immersion in any one world is always uncertain or a holding position that will surely reach the limits of its validity. Any specified domain of knowledge could be such a world. Importantly, Badiou introduces contingency to any relational structure keeping open the possibility of the currently dominant world fading into obscurity in favour of some new configuration of this multiplicity, linked again to an ontology unhampered by erstwhile conceptions of objects, relationships or priorities.

So, for example, our conception of our entire number system can be disturbed by the introduction of a new element, i, the number whose square is -1, or by Cantorian set theory conceptualising infinite sets as objects. Our examples above point to a powerful status quo that asserts traditionally understood ideas with a fixed set of relationships between them. Those ideas and relationships, however, are a function of a given world. The world of formal relations may or may not help students to enhance their more intuitive spatial awareness. Their locality might also be understood in terms of their positioning within a pedagogic world where the spatial landscape can be depicted in many diverse ways to reveal alternative configurations of objects, relationships and pedagogical priorities. For example, Williams (2012) reports on a national approach to teaching mathematics influenced by testing demands that resulted in a narrow conception of learning ill-suited to more advanced study and a reduced disposition to subsequently learning the subject. This version of mathematics, centred on mechanical application, filtered out more nuanced relationships in mathematical learning defining the interface of humans and mathematics, such as "understanding" or, as other examples, mathematical intuition, imaginative problem-solving capability, geometric awareness within bodily movement exercises, computer-mediated conceptions of mathematical fields and so forth. This feels rather like my earlier example of micro- and macroeconomics, where if you insist on doing maths in one specific way with students aged 16+, you get fewer students wanting to do it at 18+, with a composite reduction in "standards".

For Badiou, subjectivity is not centred in individual humans *qua* humans. An Internet-connected human, for example, defies all attempts to draw limits around her receptive or expressive capabilities or the control she has over them. Her individuality may be subsumed as part of a trend within big data. Badiou sees subjectivity in terms of "fidelity" to *events*. For Badiou, events comprise new ways of being in a somehow expanded multiplicity (the inclusion of *i* in number system, recognising atonal music *as* music, votes for women or an anti-slavery movement working to include more people as humans). Badiou (2009) posits alternative modes of identification with such events: one can go with it (*faithful*), deny it (*obscure*) or describe it in the terms of the old way (*reactive*). Such events disrupt the status quo triggering a wider adjustment to new conditions, consequential to a disturbance from within. De Freitas (2013) relates Badiou's notion of event to her experience of a mathematical problem that "became a problem only when it shook my cherished assumptions and set my mathematical discourse trembling with indeterminacy". The students immersed in reliving an elliptical orbit are perhaps more involved in self-reflexively exploring the apparatus through which they apprehend their spatial environment rather than the environment itself (Barad 2007). This might be seen as developing sensitivity to how space is apprehended rather than supposing that there is a correct way of doing this. Education then is not reproduction of knowledge. It is predicated on perpetual renewal, where objects, relationships and priorities persistently adjust to new conditions and to new subjectivities. Ultimately, in teaching and teacher education, we are motivated by pedagogy and productive interaction, knowing that we can never finally represent the subjects that we want to teach and educational encounters will always be about negotiating those representations.

Badiou (2007, p. 130) identifies his own conception of truth in the work of Spinoza. "Truth is the proof of itself. There is no external guarantee". Any configuration of the ellipse in "scientific knowledge" validates within that culturally specific domain of knowledge but might deplete the experience of its truth for these students, sensed as a moving circular orbit. In relation to Badiou's (2009) conception of truth, cultural or scientific knowledge never quite keeps up: "But the truth at issue, by the thrust of the real, produces a deficit in the symbolic whereby the subject, as courage, turns the radical absence of any security into its force" (p. 160). In Badiou's terms, knowledge will always need to be renewed. The three students in the previous chapter are sharing "fidelity" to an experience of truth that defies final capture in a symbolic configuration, but it also defies final capture of themselves in finished form. Their subjectivity is defined with respect to a lived experience, a moving state of affairs, not a fixed relation of object (an elliptical orbit) to subject (held by a name). Their thinking changes according to the plane upon which they find themselves. The affectivity of the space teaches the students to recognise their position or orientation in time and space through sensual clues, for example, shadows, direction of moon, darkness, temperature, foliage and reciprocities of sharing space with others. Their emergent spatial and temporal awareness marked by these qualitative features occurs as part of a layering of complex systems of relationships and spaces within constantly changing circumstances and conditions. The space that the students occupy is beyond the reach of quantifiable scientific knowledge, a knowledge that provides security and a depletion of life. The regulation that has come to prevail in many education administrations has forced aspects of mathematics into very narrow agenda governed by performance in tests. This is not always a good background for those wanting to teach, if teaching is to be any more than routine working through successive exercises. Like Roland Barthes in the prelude, they would want to give up.

In the reflective writing presented in the previous chapter, the teacher's perspective is referenced to a settled discourse, but that very settlement presupposes specific human relations to any given objects, such that words like "correct" or "systematic" can be stated, social roles can be assessed and "approval" can be granted. There is more at stake than the mere sharing of objects. It is not just reproduction of the objects but the reproduction of the world that is presupposed by their existence. The objects are linked to a conception of the wider world where social roles are set through the make-up of the world being assumed. Yet at the same time, his attempts at refusing to supply the direct answers that are sought keeps open an experimental attitude in which the final constitution of the objects and the relationships between them are postponed. After all, in this instance the exact meaning of certain terms is educationally less important than the preservation of rich social interaction shaped around the shared formation of notional objects and relationships.

The central players in this and the previous chapter have been student teachers or teachers. Their main task has been to build a language for describing mathematical experience. The teacher educators present resisted seeing the objective as being about securing standard understandings of certain concepts for onward transmission to pupils. The challenge for the teacher educators was to enable the student teachers to make up their own minds, to exercise critical capability as an attitude to mathematical learning and to build their own worlds. The latter entails their being able to articulate learning through time and to provide narratives of how ideas come into being, emphasising the experience of mathematics rather than static mathematical knowledge. In becoming teachers, the reflective engagement with how people share mathematical construction remains central, motivated as it is by the pending demands of sharing constructions with future pupils. The research of our students was centred on learning about possible relationships to mathematics in which mathematical objects and relationships were brought into *existence* rather than it being about sharing found objects, towards better understanding the educational effects that might be produced. In *becoming* teachers, they are participating in the *becoming* of mathematics. This becoming is centred on building a sense of how social interaction might work with their future students to enable the shared production of mathematical objects. Through conversation, through shared bodily movement exercises and through producing shared mathematics and reflections, the regulative discourse of the dominant order was being held at bay, until a more lively attitude had been developed enough to tolerate its arrival.

Chapter 8
Subjectivity and Cultural Adjustment: A Response to Socio-Culturalism

8.1 Introduction

I have earlier discussed how mathematics education research centred on the theme of constructivism resulted in an apparent division between researchers influenced by Piaget's individualist developmental psychology and those more in line with socially cultural models often linked to the work of Vygotsky or to sociology as a field. This chapter seeks to distinguish the theoretical model presented within this book with more familiar sociocultural approaches. The chapter revisits a debate that contrasted such an approach with an earlier presentation of my own work and seeks to take the debate forwards in the light of the work undertaken for this present book. My book *Mathematics Education and Subjectivity* (MES, Brown, T. 2011) sought to rethink:

> mathematical teaching and learning with view to changing them to meet or resist emerging demands. Through considering how teachers, students and researchers make sense of their worlds, the book explores how some linguistic and socio-cultural locations link to prevalent conceptions of mathematics education. The locations include classroom mathematics, spatial awareness, media images of mathematics, curriculum development, teacher education and mathematics education research itself. The book introduces cutting edge theories of subjectivity that trouble more familiar psychological theories of 'humans' apprehending mathematical 'concepts'. Rather, it suggests that our senses of self and of mathematics result from self-reflections within the various localities in which we live. In foregrounding subjectivity, the book shows how mathematics can provoke alternative ways of thinking towards enlivening our transformative capacities. Learning itself is depicted as participation in cultural renewal, where the very mathematics encountered is becoming something new. Addressing teachers, teacher educators and researchers, the book invites the reader to contemplate alternative trajectories of change into fresh ways of being. (Back cover)

This chapter draws on material first published as: Brown, T. (2012). Subjectivity and cultural adjustment in mathematics education: a response to Wolff-Michael Roth. *Educational Studies in Mathematics*, *80*(3), 475–490.

T. Brown, *A Contemporary Theory of Mathematics Education Research*,
https://doi.org/10.1007/978-3-030-55100-1_8

That is, we always occupy an ideologically defined location and that we might productively consider how the current state of affairs shapes our actions. Žižek (e.g. 1989) suggests that we are practically compliant to the ideologies that govern our lives even if we do not notice this compliance, since we are radicals in our thoughts and dreams. The book explored these ideological formations in mathematics education, to see how they work, so that we might see in a different way the potential trajectories of change.

The book provoked responses from a few authors. For example, one of the book's chapters, based on an earlier ESM article (Brown 2008c), comprised a critique of an ESM special issue on semiotics in mathematics education (Saenz-Ludlow and Presmeg 2006). I had sought to show how mathematics education research is often normalised according to psychological models derived from Piaget and Vygotsky, thereby shaping the form of arguments that are acceptable in the community. One of the editors of the special issue and one of the authors had shared misgivings about my approach and wrote their response in the same journal (Presmeg and Radford 2008). Meanwhile, Alexandre Pais (2015, 2016) wrote two papers for ESM critiquing my writing from the other side. He did not think that the book with its emphasis on cultural renewal was being Lacanian enough in its political reach citing "Lacan's assertion that 'the unconscious is politics' means precisely that what we think to be the innermost core of our being—the level of *desire*—is not only unconscious but schematized by politics" (Pais 2015). Pais (2016) extends my critique of socio-culturalism by showing what is lost when it reduces the Hegelian notion of dialectics to a relation between constituted entities and addresses the ongoing political failure in achieving the desired goal of "mathematics for all". Wolff-Michael Roth (2012b) detailed and interesting response to the book, also published in *Educational Studies in Mathematics*, provided a substantial appraisal in which he articulated his critique from his own sociocultural perspective. Roth provided a fairly fundamental challenge to my book in his review:

> *Mathematics Education and Subjectivity* (MES) is an important contribution: It changes us, whether we agree or disagree with it. ... I recommend MES to my readers, not because I expect them to agree with MES or with me but because I anticipate that they will come to better understand themselves and their subjectivity as they grapple with their disagreement with MES, a text designed to be controversial. (Roth 2012a, b)

It is this piece to which I will respond in detail in this chapter in an attempt to tease out some of the chief differences between sociocultural and Lacanian conceptions of mathematical learning. The wider aim of this chapter however is an attempt to make clearer the way in which perspectives derived from Lacanian analysis treat issues rather differently to more familiar theoretical formulations in mathematics education research.

As mentioned, a recurrent theme of my 2011 book was that "psychology", variously attributed to Piaget and Vygotsky, is benignly blended into theories of mathematics education research normalising certain assumptions as to how mathematics is encountered. Roth correctly reminded us that the book's ideologically defined location would create its own blind spots. In making this assertion, Roth argued that my book's account of Vygotsky was incomplete, and Roth countered the book's arguments with his own account of subjectivity derived from the work of Vygotsky. More generally, my book argued that the linguistic norms that characterise argumentation in mathematics education research result in certain forms of sense-making being overly prevalent. Consequently, certain forms of argumentation or modes of object creation available in other areas of the cultural sciences are not typically picked up by the scanners of mathematics education research leading to the exclusion of some productive approaches, including many that have emerged in the wider social sciences and humanities in recent decades. Roth argued that my book's success lay in its capacity to disrupt familiar pathways in mathematics education research but in so doing it destabilised the ground from which we could inspect newer alternatives. This opened the door to further reflect on how we conceptualise change and cultural growth through attempted reconciliation of the alternative models. Pais (2015) however objected to the book's focus on cultural renewal.

> The basic premise of Brown's approach is that culture has the power to "shape and renew the mathematics we encounter in schools" (Brown 2011, p. 167). However, for Lacan, culture functions not in the sense of renewal but in the sense of "reaction", as an ideology set to conceal some traumatic real. Could "cultural renewal" become another ideology concealing the real economic dimension of school mathematics? As pointed out by Žižek (2006b, p. 348), since dominant social systems demand perpetual reforms as a means of integrating what could be new and potential emancipatory acts into well-established social structures, the system may very well use Brown's idea of "cultural renewal" as a way to satisfy the societal demand for reforming mathematics education whilst ensuring that these ideas will not actually change any of the core features of the school system.

I do not have any particular issue with this criticism as I had seen politics being subsumed within wider culture in my definition, that is, politics shapes cultural preferences in educational practice. In line with Althusser, I do not see the possibility of stepping outside of ideology in our depictions of life. Thus, of course the assertion of "cultural renewal" is an assertion of an ideology, as the act of opening my mouth does not allow it to be anything else.

This chapter mediates some of the contrasting claims made in respect of two alternative conceptions of subjectivity. As a specific point of contention, Roth (2010a, b) aimed to "reunite" psychology and sociology through a reconceptualisation of the individual, whilst a Lacanian model includes neither "psychology" nor "sociology" in its remit. The scene is set in this chapter by contrasting how Roth and my book each reference the work of Vygotsky. From this platform, this chapter goes on to problematise the idea of the individual in relation to the two perspectives with some consideration of corporeality and of how the symbolic encounters the material. I engage with Roth's more direct discussion of Lacan towards challenging some of the other issues that he raises through showing how Lacan's later work supports persistent adjustment to new conditions. The chapter develops a Lacanian conception of subjectivity for mathematics education comprising a response to a social demand borne of an ever-changing symbolic order that defines our material constitution and our space for action. The chapter concludes by considering an attitude to the production of research objects in mathematics education research that resists the normalisation of assumptions as to how humans encounter mathematics.

8.2 On Vygotsky

Vygotsky and activity theory are discussed more extensively in Roth's 25-page response to my book than they are in the 234 pages of the book in question. I see this as indicative of how both Roth and I have strong gravitational pulls to our own preferred modes of analysis and lifetime theoretical commitments that are not dislodged so easily resulting in a disinclination to fully enter the other world. Vygotsky's theory had been introduced in my book (MES) primarily to orientate the core discussion of ideology for a general mathematics education audience. I am not a Vygotsky scholar, but as the author of MES, I was attempting to orientate the less familiar Lacanian psychoanalytic theories (e.g. 2008) in relation to better-known material in the field. Newer theories do have the disadvantage of needing to express themselves through more familiar pathways to stand a chance of being heard. This rather echoes an earlier era described in the preface where constructivism became the dominant point of reference in mathematics education research quite at variance to trends in the wider social sciences. Vygotskian-inspired activity theory, however, is Roth's intellectual home base as evidenced in numerous publications. His review spends much of its space there, critically referencing the relatively oblique discussion of the theory in MES. Notwithstanding his many complimentary observations, MES is not what Roth had wanted it to be. It seemed that he wanted me to become a Vygotsky scholar before I was qualified to speak of alternatives. His search criterion is reminiscent of the man who looks for his lost keys under a streetlight where he can see, rather than across the road where he may have dropped them. One solution might be for me to get my Vygotsky act together so that in this chapter I could more effectively counter Roth's concerns, now that Roth has recentred the debate in that domain. My preferred option however is to restore balance by emphasising that my own long-held core frame of reference is centred on the Lacanian theories of Badiou and Žižek who continue writing to this day with no reference to the Russian and his followers as far as I know. I refer many more times to these authors, authors that Roth does not mention in his review and in so doing Roth misrepresents the main theoretical thrust of the book. The book is centred on showing how contemporary theory by living writers offers new analytical resources. In restoring balance, however, the chapter will keep to the areas of concern that Roth shares whilst resisting his tendency to see the issues exclusively through his chosen analytical filter.

Given Roth's chosen theme, it is surprising that his review so quickly skates over the most extensive direct comparison that MES makes of Vygotsky and Lacan. As MES indicates, Lacan and Vygotsky would both claim that humans feed off the linguistic apparatus that surrounds them. For both authors, "We become ourselves through others" (Vygotsky, cited by Roth, all Roth quotes are from his paper). They would wholly differ, however, in their understanding of how humans and their formation relate to symbolic mediation (more later). Vygotsky's notion of zone of proximal development (ZPD) has been popularised in many instances of mathematics education research as bringing children into the social world. I trust that Roth is more precise: "through the child, the societal becomes individualised and concretised". Yet Vygotsky's work in the very different circumstan(encountered during his lifetime a century ago has been subject to multi' within the cultural imaginary of mathematics education research. Bibby (p. 38) argues that the "seductive imagery conjured by Vygotsky's metaphor ... allows us to ignore the difficulties and resistances which the learner will encounter and develop". She continues: "the metaphor encourages us to ignore any differences between the learner and the teacher and seems to suggest that the learner's differences will be unimportant and willingly subjugated to the teacher's benevolent intentions".

A number of authors have considered the notion of alienation in the mathematics education literature (e.g. Williams 2015; Radford 2016). Whatever depiction we choose, Vygotsky's ZPD contrasts sharply with Lacan's (1986, pp. 203–215) assertion that humans' alienation from language is built into their very constitution as subjects. As we shall see later, the subject's constitution in Lacan's formulation is *not*, as Roth persistently suggests throughout his piece, divorced from the body or living being. For Lacan, however, the language used to describe people never quite fits with their own sense of reality, "the imaginary is enough to motivate all sorts of behaviour in the living being" (p. 207). And they can be alienated from the very apparatus used to include them. In Vygotsky's model, the child's environment provides both the form and content of his personality, even if that personality is "individualised". On the contrary, for Lacan, dialogue functions as the alienating experience. Teachers may or may not identify with specific aspects of the curriculum they are charged to present. Children may or may not connect with the account of the world that the teacher provides. The space between the place assigned and the place taken results in a "permanent hunger" to close the gap (Emerson 1983). This hunger is never satisfied. The only way out of any restrictive caricature of self is to accept the turbulence of participation in discursive activity, and this participation produces real effects on the body's formation, not all accounted for in any accompanying narrative. For Lacan, any attempted identification with specific discourses or ideologies is tainted by the individual's desire to please and to respond to the demands she perceives (from the Other), even though, as Lacan claims in his later work, those demands may not actually exist. Importantly, however, the difficulty in fit, the alienation, can be experienced as a positive condition, releasing an individual who has grown out of the discursive clothing bestowed upon her.

Roth contends: "It is evident that the Russian scholar has anticipated Lacan". I am rather sceptical on this point. Roth overreaches himself in attributing rather too much of the thought of the late twentieth century, and especially too many aspects of Lacan's writing, to being a later-day exemplification of Vygotsky. Lacan is firmly rooted in Freud and Hegel who both predated Vygotsky and wrote much more. It may be that Vygotsky provides one solution to the issues in question, but the point of MES was to show how Lacan offers an alternative approach. Lacan's work as developed by more recent writers better supports more recent conceptions of subjectivity introduced long after Vygotsky passed away and which provide an alternative to present-day Vygotskian formulations.

In the next section, I seek to paint the new territory occupied by Lacan and link it to the work by Žižek and Badiou. Conceptions of "psychology" as attributed to Piaget and Vygotsky, so often used in support of mathematics education research, take an altogether more marginal place in MES, as strictly alternative points of reference. The wider notion of subjectivity shifts the focus of the book onto the multiplicity of readings available in the diverse circumstances we face today where consensus on how the world is marked out is not readily achieved. The generation of theory provides alternative analytical filters through which we can read contemporary circumstances, as exemplified in one of the MECT Special Issues of ESM (Brown and Walshaw 2012).

The remainder of the chapter addresses Roth's discussion of Lacan within MES. I commence with a brief sketch of Lacan and two of his followers.

8.3 On Lacan

8.3.1 Lacan, Žižek and Badiou

Lacan's notion of the subject was initiated through his work in psychoanalysis with individual clients. The accounts provided by these clients became the material for his analysis. These accounts comprise localised cases of the wider discursive network, a revelation that loosens their connection to the client seen as a stand-alone living being. (This is not to say that the living being was unaffected by the production of these accounts – more later.) The human subject was defined according to the descriptions available within this network. Indeed, the accounts alerted us to how human individuals derived from this wider network. Individuals might no longer be considered primarily as stand-alone biological entities but rather as consequences of chance events, or social movements, where the individual is understood in terms of his or her identification with these events. The Internet, for example, produces conceptions of humans. Facebook can celebrate the personalities of individuals but then convert them into mere statistics in a large-scale consumer survey linked to a sales drive or election campaign.

Many perspectives on Lacan present in MES have been accessed through Žižek and Badiou, who are each both major thinkers, explicitly responsive to contemporary cultural and political themes.

Žižek's work is centred on how culture (films, artistic productions, jokes, flower arrangements, news reports, television broadcasts, the Internet, PISA test items) is revelatory of how the society thinks of itself. Cultural life is not so much centred in the individual. Rather, the individual is understood through his or her identifications with or participation in certain aspects of cultural life. Yet, in this Lacanian formulation, these identifications are never quite secure. The subject mistakenly recognises versions of herself in this symbolic network that are never quite sustainable. Try as I might, I am not like George Clooney. This alienation, the gap between place assumed and the place assigned, mobilises subjectivity to find a more comfortable space, yet instead finds that it cannot be encapsulated in any given symbolic form. No story quite fits. Life in such circumstances is governed by unconscious forces and set moves, which shore up the gaps in any overt story that an individual might confidently present.

Badiou's notion of subjectivity (e.g. 2009) also takes a radical step beyond a concern with the individual human in a therapeutic encounter. He drops any privileged link to the living being in favour of seeing subjectivity in terms of identification with a movement to a new state of affairs. For example, Spartacus was instrumental in an anti-slavery movement that transcended the individual human Spartacus. Spartacus' identification with the anti-slavery movement, the collective assertion of a cause, was more important in locating subjectivity than his individual humanity as a biological entity. Thus, subjectivity is associated with a redistribution of the psychological, where perhaps our whole concept of what it is to be human (a teacher, a student) has shifted to a new configuration, and where perhaps the individual human's operative role is rather less central than was previously supposed. Critchley (2008, p. 44) argues: "One can only speak of the subject in Badiou as a subject-in-becoming insofar as it shapes itself in relation to the demand apprehended in a situation".

8.3.2 The Place of Subjectivity: The Case of the Mathematics Education Researcher

In addressing the term "subjectivity", one may reflect on one's own common usage of other familiar terms (such as individuality, sociality and psychology). Roth introduces Leont'ev's activity theory towards criticising what he sees as overly casual use of the term "social" in my book.[1] He distinguishes the term from "societal", which he sees as relating to the political/ideological system. It seems unproductive to spend too long differentiating between the ways in which the terms are used by alternative traditions. My point had been to contrast Radford's teaching approach with an alternative approach in pinpointing subjective engagement. The difference related to the way in which the terms of reference for the given activity (or language game) were set and whether these terms were negotiable or not. The students either followed sequences predetermined by their teacher in Radford's example or set their own parameters for sequences in the activities described in my book. The demands from one case to the next were very different. The student response was a function of how he or she was subjected to the pedagogical space in question.

As mentioned, I addressed the issue of subjectivity in my response to an *Educational Studies in Mathematics* special issue on semiotics, which led to a reply from authors involved (Presmeg and Radford 2008). The issue at stake also

---

[1] The reviewer of the current manuscript advised me to leave the word "social" out of the title. She/he was right. And thank you for your very careful reading and your excellent ideas.

related to how individuals respond to a given field for action. I conceptualised the subject "mathematics education researcher." What is demanded of such a designation (journal or funding agency criteria, employer expectations, professional self-image, etc.)? How do individuals follow such a designation? Are there preferred ways of aligning with the designation? The mathematics education researcher could research how to improve the current set of teachers (by improving their techniques, changing the curriculum, setting new priorities), or she/he could research how to get a new set of teachers (paying people to train in this area, relocating troops into teaching, benefitting from the new popularity of physics). Do, for example, prevalent conceptualisations of what research is lead to a disproportionate number of research papers where certain perspectives are revealed, thereby normalising particular accounts of what it is to be such a researcher and in turn what constitutes research? For example, as seen, much contemporary research in mathematics education takes its starting point as responding to the demands of TIMSS or PISA, whether that be complying with those demands by showing how TIMSS results can be improved, or by critically resisting that formulation of mathematics. Lacan conceptualises subjectivity more generally as being a response to a demand or an expectation of what is required by a given designation. I am still working on Zizek's unorientable alternatives (2020).

Authors in the special issue discussed a range of themes, but MES argued that the emphasis of the work overall supported the proliferation and normalisation of familiar research perspectives. In the case in question, there was a tendency towards using Piagetian and Vygotskian psychological models as though these models were an essential component in any historicisation of development in mathematics education research. That is, the subjectivity of "mathematics education researcher" was conceptualised with respect to well-known psychological filters. More generally, MES sought to argue that a disproportionate volume of research in mathematics education is directed to the improvement of teacher technique according to a given regime, perhaps at the expense of ignoring other more effective levers. Similarly, Roth's account of a teacher-student dialogue emphasises the quality of interpersonal exchange, within a rather localised activity framework. The need to meet publishing criteria can influence the research author's conception of who they are and what they are trying to do, the style of paper submitted and the way in which mathematics (e.g. seen as knowledge, analytical apparatus, problem-solving or basic skills), teachers (e.g. as didacticians, facilitators, inspirational figures, carers) and students (e.g. performing in tests, independent thinking, obedient) may be conceptualised. Roth asked the question: "Do we tell what has happened to us during any particular working day in exactly the same way to our 5-year-old son, our mathematics education colleague, the hairdresser or spouse? We don't!" There is however a risk that we always go down the same tram tracks when talking to our audience of mathematics education colleagues since our working environment is governed by certain norms, preferences, habits and expectations, which result in certain styles of familiar action that may preserve past inequities, redundant models of practice and tired theoretical paradigms. MES argued that there are substantial gaps in the scope of mathematics education research, which is not the fault of individual authors as emphasised, but rather the economy of such research does not support interest or coverage in certain areas. There are blind spots. The ideological dimensions of mathematics education shape practices, practices about which we are not always fully aware. We must persistently attend to the assumptions that we are making in setting the terms of reference for mathematics education research.

8.3.3 Language Games and Renewal

Individuals need to employ mainstream styles of communication if they are to participate in mainstream life. Examples were mentioned above in relation to mathematics education research as a discourse. Compromises need to be made towards mainstream discourse if one is to participate, whether that be towards constructivism or Vygotsky's socio-culturalism. Roth's critique notes two places where MES "complains" that the "individual is obliged to use these languages if they are to be included in social exchanges" (MES, p. 105); "[i]n this way the human subject identifies with something outside of himself. They see themselves in the social languages, but the languages never quite fit" (ibid). Roth associates these two statements with participation in language games as if the games already exist and can be participated in according to certain fixed rules, such as in a game of football. As suggested above, however, identifications with the discursive environment in the open sense that Roth depicts by way of Derrida are never quite secure. I fully applaud Roth's opening remarks where he states: "With every word, (the old) language dies and (a new) language is re/born". This is a point of strong agreement between us. Although Roth seems to be doubting this point when its sense shifts later: "It may be detrimental to good theory if the categories shift in translation". Similarly, the theory may slip if meaning shifts in translation from one use (Wittgenstein) to another. In Lacan's conception of the subject, however, the deluded fellow mistakenly recognises and lives by versions of self in these symbolic networks that are not sustainable. That's maybe why he gets depressed. The storytelling individual cannot keep up with events and casts an imaginary layer over everything to make sense of the turbulence in unpredictable ways. The gap between the place assumed and the place assigned mobilises subjectivity such that it cannot be encapsulated in any given symbolic form. It is this very failure that gives the subject license. In the first statement, rather than complaining, MES was hinting at the costs and benefits associated with fitting in with the current collective story. This is rather akin to Roth's statement about a teacher (Mrs Turner) talking to a pupil: "The language, however, is not that of Mrs. Turner. She does not invent it here, but it has come to her from the generalized other, to whom, in her utterance, it returns. She is not only the subject who uses the language, but she also is subject to it and the things it can express." The second statement slightly disrupts this however. MES was celebrating the human subject's ability to transform the state of affairs because of the rules, or the language, never quite working. The alienation can be experienced as a positive condition that renews the conception of the "game" guiding action. There is not a game as such, but rather successive shifts of discursive filters that can successively and radically redefine the field of play (for the game, as it were).

In the case of algebra, for example, I do have the option of playing to the rules of established school algebra, its familiar forms, procedures, etc. But can I be sure that those rules really are stable? If I was to consult one of my son's school exam papers in this area, I would find a much-depleted conception of algebra propping up a test designed to be consistent in style with wider TIMSS/PISA assessment where algebraic concepts are partitioned in very specific ways into questions of a chosen or familiar form. Quite apart from the formal rules of algebra, that which counts as school mathematics is constantly shifting since the pedagogical/curricula layers are permanently on the move in response to ever-shifting administrative demands (Brown and Clarke 2013; Tang et al. 2012). More positively, there will be wholly contemporary depictions of geometry, such as those developed within recent technologies, opening whole new worlds of spatial awareness. Geometry is not independent of its social filter or language game, except that we are in a permanent state of adjusting to the supposed rules of new games, or new emphases, adjustments triggered by failures of fit within previous versions of life. Any such cultural adjustment needs to be worked through by individuals and by groups of individuals who are never in the place of their ancestors. "With every word…".

8.3.4 Corporeality and the Real

As seen Lacan's psychoanalytic procedures produced accounts from patients as symbolic material, derived from wider discursive activity – an example of how people talked more generally. As a patient, this would make up part of the story of who I am and of who we are. But this story also produces who I am as a physical entity in tune with my environment. For example, within mathematics education research, there has been much work on the theme of gesture and with how mathematical phenomena are referenced or evoked by bodily movement (e.g. de Freitas and Sinclair 2014). This work might be understood as an attempt to understand the subject's identification to the physical world as seen through a mathematical lens. Mathematical understanding is thus expressed through gesture. The more general issue, however, relates to how the subject connects with the world through a mathematical or scientific lens. How does the subject produce herself within a world understood mathematically or scientifically? MES, as in the last two chapters of this present book, provides extensive discussion of students bodily situating themselves in, or moving within, large spatial environments as understood through certain mathematical or scientific filters, jokingly referred to as "extreme gesturing". Pedagogical apparatus more generally however is produced according to supposed modes of apprehension, such as inside/outside, within a count, grouped according to criteria, in the form of a graph, having been shrunk to an infinitely small point, etc. A mathematical account perhaps comprises the endpoint of a process of achieving an ever more precise story of my experience, such as in reaching a generalisation. I am the subject of the story I tell and reveal who I think I am through the way I reflexively situate myself in the telling of that story: a portrayal of a mathematical me. Žižek's work is centred on the fact that we declare who we are through our cultural productions. Likewise, we might assert our collective mathematical identity, or more specifically, what counts as mathematics in schools.

In Lacan's (2008, p. 81) terminology, this storytelling might be understood as follows:

> The subject is dependent on the articulated chain represented by science's acquired knowledge. The subject has to take his place there, situate himself as best he can in the implications of that chain. He constantly has to revise all the little intuitive representations he has come up with, and which becomes part of the world, and even the so-called intuitive categories. He's always having to make some improvements to the apparatus, just to find somewhere to live. It's a wonder he hasn't been kicked out of the system by now. And that is in fact the goal of the system. In other words, the system fails. That is why the subject lasts. (quoted by Brown, T. 2011, p. 123)

In other words, the scientifically defined universe contingently defines worlds (e.g. Euclidean conceived space, food security patterns, gross domestic products) and the physical coordinates of the human's place within them. The individual (such as the child described by Roth living on a coffee plantation), however, may not be especially comfortable with these assigned places provoking consequences to these perceived failures of fit (e.g. medicinal, nutritional, statistical, normative). For example, Piagetian psychology, so influential in earlier accounts of mathematical learning, defines individuals in terms of various physical or responsive attributes, or developmental stages, which may bypass the affective or creative sense of self possessed by the individual herself. Or alternatively, the individual human might too compliantly accept externally applied designations, e.g. economic rather than epistemic and social welfare – a reduction of life that will ultimately be resisted.[2] Lacan's model locates life as a negotiation in which the individual works through successive accounts of the world, each of which points to a place for the individual. Lacan mocks the failure of scientific constructs to keep up to date, consigned as they are to the need for regular renewal, whilst the human always survives. For example, economic models are notoriously unstable yet maintain a crucial presence in our attempts to control our relation to life through mathematical apparatus. Physical models of the universe move rather more slowly, but no less radically. But what lies beyond this symbolic modelling? Or perhaps, how is the modelling motivated? Lacan's answer is "the Real". I need to clear some preliminary points, however, before explaining this important term that is missed in Roth's analysis.

Lacan always moved on, defying any straightforward representation of his ideas. That was his point; ideas are never stable in relation to the world they seek to depict or in relation to the people having those ideas. One only needs to read any random paragraph from his immense body of work, or the three paragraphs included in this

[2] This resistance would take the form of *jouissance*, a surplus to the discursive experience.

present chapter, to realise he favoured a poetic style and the provocation of unsteady responses over the delivery of stable ideas. Notwithstanding Roth's greater access to Lacan's style through his linguistic background, Roth's review (e.g. 3.1, 4) focuses on controversial readings of a very specific phase of Lacan's work,[3] namely, the middle period from the fifties, where the influence of Saussure's structural linguistics was at its greatest. In addressing this aspect, Roth incorrectly separates Lacan's three orders of the Imaginary, the Symbolic and the Real, which relate symbolic activity to the tangibility of the world we encounter. The work of Žižek and Badiou referred to in MES is centred on a later Lacan. By this time Lacan had been exposed to a more diverse audience stretching well beyond the therapeutic community, in seminars hosted by the Marxist philosopher Althusser (Tomšič and Zevnik 2016, p. 4).

The key difference between the middle and later periods of Lacan's work is the prominence in his later work of what he calls "the Real". The Real is variously defined by him over the years but relates to that which is beyond the scope of representation, "that which resists, the impossible, that which always comes back the same place, the limit of all symbolisation" (Lacan, quoted by Critchley 2008, p. 63). Critchley continues: "The basic thought here is that the real is that which exceeds and resists the subject's powers of conceptualisation or the reach of its criteria". The thought can never comprise a well-defined signifier or signified. Lacan's later emphasis on the Real cuts across Roth's supposition that "Lacan focuses exclusively on language". For Lacan (1986, p. 221), "philosophical idealism … cannot be sustained and never has been radically sustained." Badiou or Žižek assigns Lacan's work to wholly materialist projects. The Imaginary,[4] the Symbolic and the Real, key terms in Lacan's apparatus, famously comprise a Borromean knot of mutual dependency. In MES, the Real underpins the mechanisms for change that are depicted, where the Symbolic perpetually chases a Real that defies any final encapsulation.

The Real itself can be responsive to, or be altered by, these attempts at its capture. The physical state of clinical depression can be improved or worsened by talking about it. Similarly, bodily intuitive conceptions of space, such as Roth's example of feeling a cube, or examples in MES that "in the limit come close to the idea of a circle" (Roth), can be transformed through introducing novel ways of talking about our spatial movements. Roth's (3.1) suggestion that "Lacan never was concerned with real material life but only with the accounts his clients provided thereof" is inaccurate. It is not an adequate representation of the pain experienced by his patients or of the management of that pain by the analyst. The misery was all too real. Lacan (1986, p. 203) insists that it "is the field of the living being in which the subject has to appear". The physical state of a body, including its feelings, is a function of how it is mapped out or classified by medical experts, which in turn has an

---

[3] I have not followed Roth in reading Lacan in the original French, even though Derrida and Lacan, alas, never quite reached a final resolution on each other's obscure texts, despite both of them being French.

[4] Lacan's iconic example is of a young child looking into a mirror and recognising the image as herself, an image that suggests a completeness that may not be experienced.

effect on the subject's own awareness of her physical make-up and how she is quantified for medical assessment. The patient may develop awareness of her own bodily condition and how she adjusts various medications to produce particular states of physical well-being. Similarly, exercise programmes are quantified (reps, resistances, speeds, weights, timings, targets) and may be adjusted to produce different effects on the body. Likewise, the immersion of students in spatial environments (e.g. How do I experience moving on a really big circular locus?) works on the students' physical sense of self ("the force overcoming the resistance of the body to walking, the opposition of the body to gravity or the walking of the walking" (Roth)) rather than just generating mere reportage of that experience. As in gesturing, the movements and sensations are part of how they learn mathematics. This negotiation, however, whilst peripherally aware of the Real, can never directly represent it. "My knowledge of myself is limited to the empirical presentations that pass before my gaze. What I am - ontologically - remains a gap in knowledge. In Lacanian terms, we are only ever presented with imaginary egos and subjects of statements, but never the subject of enunciation" ("The accursed share", anonymous blog).

8.3.5 Discourse, Relationality and Subjectivity

"Words do not belong to one person, but constitute the realities for two; words are not the words of individuals, but always belong to speaker and audience simultaneously". Roth attributes this sentiment to both Derrida and Lacan. Yet surely this image of two people talking is locked into conceptions of a circumscribed individual (a subject of psychology) alien to both writers. Derrida did not spend much of his time talking about individuals or reality. Lacan's work was entirely about subjectivity but where the psychologically defined individual is less prominent as a distinct entity. Rather the subject is understood relationally in terms of his or her identifications with particular aspects of life, such that it becomes unclear where the individual ends and the world begins. Lacan totally rejected ego psychology's project. Roth's inclusion of the transcript reporting on a conversation between Mrs. Turner, Mrs. Winter and Thomas provides a typical example of how he sees individuals interacting on mathematical tasks, where, for example, Mrs. Turner is "allowing Thomas to understand (the meaning of?) the question". It seems reasonably straightforward to decide where, as individuals, Mrs. Turner ends and Thomas begins, even if they share "realities". Roth has written many other such papers where the expressive physical gestures of the individual human extend beyond the sharing of spoken or written symbols. He has also responded to discussion in MES where students experience walking the loci of various geometric configurations. There are also many instances where the student's demonstration of his or her mathematical understanding amounts to (or subjectivity is reduced to) little more than filling in a gap in a story provided by some sort of assessment device.

Psychology and sociology each have long histories as academic disciplines, but not that long, maybe 100 years or so. They each comprise a way of looking at the world where certain entities are privileged in their analyses. There is a difference of emphasis between the ways in which Roth and I are each centred in conceptualising subjectivity in relation to psychology and sociology. In his broader project, Roth (2010b) aims to "reunite" psychology and sociology. He focuses on the individual human individuating the collective programme through his or her expressive action, such as in an exchange between teacher and student. More typically, MES focuses on how discourses shape subjective action within a Lacanian model that includes neither "psychology" nor "sociology" in its vocabulary. Students were asked to report on their memories of learning calculus at school. Teachers were asked to reveal their agency in implementing new curriculum materials. The work of researchers in mathematics education was analysed to see how the work encapsulated the field. That is, MES (p. 129) asks: "What aspect of the whole person is activated (or brought into being) in any given semiotic configuration?"[5] How are they created as subjects? Which discursive aspect responds, or appears, and why?

MES (p. 127) consults Lacan on this point who writes in his usual playful manner:

> The whole ambiguity of the sign derives from the fact that it represents something for someone. This someone may be many things, it may be the entire universe, in as much as we have known for sometime that information circulates in it... Any node in which signs are concentrated, in so far as they represent something, may be taken for a some-one. What must be stressed at the outset is that a signifier is that which represents a subject for another signifier. (Lacan 1986, p. 207)

The "ambiguity" for Lacan is centred on how the "someone" is predicated in semiotic activity. What does Lacan intend by his curious suggestion that the "someone" could be the "entire universe". This term is made yet more obscure by the clause "in as much as we have known for some time that information circulates in it". This hints at a more extensive engagement with discursive networks and their production of subjectivity, a subjectivity that can never quite hold on to the discursive universe that it reflects. Connectivity to the Internet, for instance, recentres our sense of self, our sense of reach and our scope of receptivity. It affects how we process information, make gestures, impact on others, etc. MES and the present book address how teachers, students and mathematics itself are commodified according to the needs of an exchange economy. Contemporary understandings of subjectivity centred on human immersion in discursive and signifying activity provide a backdrop to Lacan's pre-Internet assertion that "someone" might provide access to the entire network of discursive activity. Everyone is implicated in the discursive construction of society and everyone draws on that construction. And thus: "Any node in which signs are concentrated, in so far as they represent something, may be taken for a some-one". A subject then is not just an individual human, but it could also be an agency, a cause, a movement or a "fidelity" to a new way of being (more on this shortly). The final sentence in Lacan's paragraph "that a signifier is that which represents a subject for another signifier" might be related to an example referred to in MES:

[5] The rather troubled notion of the "whole person" must have slipped into the text accidentally.

> The old-style hospital bed has at its feet, out of the patient's sight, a small display board on which different charts and documents are stuck specifying the patient's temperature, blood pressure, medicaments, and so on. This display represents the patient - for whom? Not simply and directly for other subjects (say, for the nurses and doctors who regularly check this panel), but primarily for other signifiers, for the symbolic network of medical knowledge in which the data on the panel have to be inserted in order to obtain their meaning. One can easily imagine a computerised system where the reading of the data on the panel proceeds automatically, so that what the doctor obtains and reads are not these data but directly the conclusions that, according to the system of medical knowledge, follow from these and other data. (Žižek 1998, p. 74)

The signifier, a graph maybe, represents the subject, a patient in the bed, for another signifier, a doctor or nurse reading the graph with view to it impacting on a specific dimension of their subsequent actions. That is, we are not attending to patient or medic as "whole people". Rather we are considering the patient through the restricted registers of the patient, with particular symptoms, and a medic only interested in those symptoms (perhaps with view to setting a correct dosage), according to the wider system of medical knowledge. One could extend the computerised system so that a sensor could detect a bodily change that triggered some medication being introduced into the bloodstream.

This example echoes countless studies in mathematics education research where there is a demand to isolate the mathematical dimension of wider discussion, but that very isolation serves to reduce the lived context. Such questions are crucially linked to the geography of the supposed interface of subject and object. Mathematics in schools exists substantially as pedagogical material crafted for supposed modes of apprehension. Students are required to spot certain things according to the given mark scheme. But such apprehension depends on how we understand mathematical objects and how we understand human subjects. That is, a given mark scheme supposes a given conception of a student able to answer on those terms, and supposes that mathematics can be seen in a definite way, and taught by a teacher able to evoke it in that way. That is, as above, subjectivity is reduced to little more than filling in a gap in a story provided by some sort of assessment device. In another example above, I queried how the subject "mathematics education researcher" derived from the demands placed on that designation. Roth and I have chosen different terms of reference in making this assessment.

8.3.6 Subjectivity, Relationality and Personality

In a similar vein, I fully agree with Roth when he says that "we cannot stop with our consideration of the subject and subjectivity by considering what happens in a mathematics classroom alone. ... A person cannot ever be identified by its subjectivity within the mathematics classroom or within a mathematics education discourse". Subjectivity cannot be partitioned into just those bits concerned with mathematical learning. Seeing mathematics education as so many classrooms organising mathematical learning is only one version of how mathematical learning is taking place in the world today. The subject, or the "human", or the "personality", in Lacanian terms derives from persistent (failed) attempts to make sense of the world. We can never get our story quite right. The Real can never quite be captured in the Symbolic, even in a given subdomain of that Symbolic, such as that relating to the mathematics classroom. Lacan's subject (of desire) is always reaching beyond the current state of affairs, a perpetual quest to improve on the current story motivated by spotting the "holes in discourse" (Lacan 2008, p. 27).

In the hands of Badiou or Žižek, Lacan's motivation entails detecting the limits and limitations of the ideological parameters that shape our actions. Badiou's work, for instance, is centred on the potentialities of noticing blind spots in our current story and how these blind spots might alert us to new perspectives, to new ways of being. Any world relates to a state of knowledge. Knowledge, however, does not capture Truth (for all), and for this reason, knowledge will always need to be revised to fit the times and circumstances. For example, mathematics (as knowledge) was expanded when Cantorian set theory permitted infinite sets to be conceptualised as objects, and again when the real number system sought to include i, the number whose square is -1. For Badiou, there is some mathematics that is a function of contingent empirical reference (e.g. mappings of phenomena observed in the physical world as we presently know it) and some that is not dependent on such reference (Badiou uses set theory to create his model.) But we occasionally have to shift ground as we are not always entirely sure as to how much mathematics is motivated by some reference to a world. Indeed, mathematical thinking relies on shifts of attention (e.g. Mason 1989) to differentiate between particular and general dimensions.

Perhaps as an example of this differentiation, Roth (3.3) cites Rancière for whom subjectification denotes "the production—through a series of action of a body and of a capacity for enunciation not previously identifiable within a given field of experience, whose identification is thus part of the reconfiguration of the field of experience". I take this to mean that a hole in discourse has been located and that a necessary adjustment has been carried out. Roth rephrases this as: "the subjects are transformed by their own actions that are themselves a function of the field of experience and therefore are not entirely owned by the subject". It seems to me however that the two authors are using the terms "subject" and "body" differently. Rancière, I believe, is using the term "subject" in much the same way as Badiou as described above, whilst Roth is seeing "subject" as being linked to an individual human body. Roth is incorrectly assuming that Rancière is also referring primarily to an individual human body. In the work of Badiou and Rancière (thinkers who I have witnessed sharing the same stage), I suggest, we are witnessing a radical redistribution of the psychological where, within Badiou's Maoist preferences, individual personalities follow from a more collectivised account of the world. That is, individuals follow communities of practice adjusting to new ways of living.

In Badiou's terminology, bodies (whether that be an individual body, or a collective movement, or a body of thought) may be understood in terms of subjective "fidelity" to specific cultural adjustments, that is, to events, which comprise new ways of being in a somehow expanded multiplicity of elements (the anti-slavery movement working to include more people as humans, or votes for women doing the same to expand conceptions of the electorate and of democracy). One might also consider changing university entrance requirements to rewrite the conception of a graduate to meet new workplace criteria, highlighting new pedagogical/mathematical objects/priorities consequential to the growing influence of international comparative testing.

The domain of subjectivity is activated and renewed by such events and hence the possibilities of what it might be to be human. That is, we are not just concerned with humans changing the material conditions as Roth suggests but also changing the conditions through which it is understood what it is to be human or more specifically what it is to be a teacher or a student. For example, so many "human" interactions are now processed through technical media, affecting spatial and temporal parameters, and thus how subjectivity is produced, represented or accounted for. The living being is sometimes less prominent in this virtual landscape than in the exchange Roth describes between Mrs. Turner, Mrs. Winter and the pupil Thomas. Roth's suggestion that "Thomas's own utterance is an integral part of the production of the subject" implies a singular subject "Thomas" in just one place, with the rather flat suggestion that Lacan sees "the subject in the relation between the signifiers", as if some formula of identifications could produce a personality or a clearly defined sequence of subject positions.[6] Thomas, however, has different ways of occupying the space. Lacan (1986, p. 208) argues that the subject "develops its networks, its chains, its history, at an indeterminate place" beneath the signifier, or dominant story. The "subject may in effect occupy various places, depending on whether one places him under one or other of these signifiers" (p. 209). These multiple opportunities to set the coordinates defy stability or consistency in perspectives or descriptions since the perspectives comprise the learning of new ways of being that might transcend the immediate physical territory of the three people present, such as in following the wider introduction of a new mathematics curriculum, in aligning with a new attitude to curriculums or in working practices adjusting to new systems or technologies (Hoyles et al. 2010).

8.4 Conclusion

The method in psychoanalysis entails the production and analysis of symbolic material, or of a story. In mathematics education research, we need to attend to the texture of what we produce. The story in itself is a valuable entity, which methodologically produces the research objects that orient the mode of enquiry. This story is not subservient to something that it is trying to represent (such as how a mind works, or how ideas have been portrayed through the work of historically signifi-

[6] Lacan's subject was "barred", as in Roth's Fig. 2, to emphasise the gap between the subject's place of enunciation and the enunciated subject. There is a difference between the individual and the way that individual implies herself through her descriptions of the world. Similarly, in naming my son Elliot, there is a gap between how I visualised that name and how Elliot now lives it.

cant writers, how a meaning has been fixed, or usage familiarised). Indeed, the story is productive of that thing and a useful barometer of that thing. It entails looking at one's own looking to see how objects (meanings) are generated within a story that never settles. But the story is also productive of the person telling the story, since the story reflexively situates its storyteller. We must, however, be cautious. Lacan suggests that when the analysand says "I", the analyst should be mistrustful. In responding to Roth, it is more precise when "I", like Roth, refer to the MES text rather than speaking in the first person as the author. The individual, or any collective, is only ever partially self-aware. "I" am surprised by some of Roth's claims as to what MES is saying, whilst learning a lot through that surprise, and for which I am very appreciative. The stories we tell are both part of ongoing speech and part of the wider discursive network. The location of the stories will always move on since speech never stands still. They have a limited shelf life. The stories will adjust to new circumstances. And it may be that our storytelling resources will change, such that we tell stories in new ways to produce alternative effects. Different stories will be told before long. But it is possible to learn from these present efforts. That is, we can learn from how those attempts fail to produce the result that we seek. Persistent attempts produce patterns of failure that allude to the Real that is sought. For no part of the Real is there a final encapsulation. It is only ever possible to begin with past illusions, or localised predictabilities. Any adjustment adjusts the whole picture, not just some localised elements. There is no progress through a tick list of certainties.

For Lacan (2008, p. 17), "truth is always new", and knowledge is always renewable. But that knowledge provides much of our everyday reality. The emphasis on the stories that we tell is not to suggest that we reject the knowledge that we have. We can learn from how those imposed stabilities guide life, or sometimes, whole lives. We may assume specific discursive formations, set rules, introduce analytical frameworks or hold certain assumptions for the time being, which influence the research questions that we ask. For such knowledge is a function of the worlds in which we live. Indeed, much of our infrastructure (buildings, modes of governance, law, social practices, preferred styles or pedagogical objects, curriculum forms, schools, conceptions of teacher, examinations) is a function or reification of how previous generations conceptualised life. We can however better appreciate the limits and limitations of such worlds and the forms of knowledge that they host, to avoid the false comfort in contingent arrangements and to better understand how those arrangements shape our actions. As in many instances of life, we are swayed by our own preferred versions of common sense, and these influence the research that we pursue. This piece of writing is arguing that we might learn more about our own common sense to better understand its effect on our lives.

The task of research surely is to generate alternative arguments, not to suppose that there is a neutral scale that allows us to cross-evaluate, or mediate, or maximise. The purpose of this chapter has been to argue for theory, not so much for a particular type. And theory moves on in response to changing circumstances. To reference everything back to old writers can trap our thinking into the false security of established modes of thought and their priorities that can fix both objects and the relationships between them. My book *Mathematics Education and Subjectivity* explored how different sorts of common sense are revealed in instances of mathematics education practices and in the discussions that surround this type of education. That book was concerned with showing how we might work against those forms of common sense that prevent us moving to fresh ways of being that might serve us better in new circumstances. In that quest, Roth and I are certainly at one. In MES and in this present book however, I have sought to let the Lacanian model that I want to assert to stand more on its own.

Chapter 9
The Evolution of Mathematics

Mathematics has maintained an enduring image as a field of knowledge that lends its resources to many intellectual pursuits and practical applications. School mathematics, however, has responded to a commonly conceived purpose of supplying the world's workforce with the resources needed to support economic well-being. This motivation has perhaps superseded earlier educational priorities centred on social welfare or epistemic motivation. Research intended to inform the practices of mathematics classrooms has often reflected local interpretations of this fundamentally economic agenda. Since the advent of international comparisons, for example, governments have been jockeying for a better position in the resulting league tables. Good performance in international testing programmes has been interpreted as indicating wider economic competitiveness. Relatively poor performance, however, has often been cited to justify changing educational policies. For example, we have seen earlier how a British government policy statement for education in England proposed to expand employment-based models of initial teacher training. It explicitly cited England's performance in international comparisons such as TIMSS and PISA as a reason for pursuing this approach to training, so that children will compare more favourably with their peers overseas (DfE 2010). The (dubious) rationale of the document was that by enabling "a larger proportion of trainees to learn on the job" and by "learning from our best teachers" (p. 23), student teachers would more effectively encounter the realities of school and be better able to implement centralised curriculum and assessment, which are in turn designed to improve England's international performance.

In this concluding chapter, I argue that our conceptions of mathematics and of ourselves as researchers, teacher educators, teachers and students move on through a broad range of pedagogical or practical agendas, such as improved economic competitiveness or performance in international comparisons. I seek to offer a more theoretical account of how the evolution of mathematics more generally might be understood in terms of cultural or political adjustment, or according to popular or official conceptions of what they should be. That is, we propose that the empirical

T. Brown, *A Contemporary Theory of Mathematics Education Research*,
https://doi.org/10.1007/978-3-030-55100-1_9

reality of mathematics today, the everyday practical business of mathematical activity in many locations, feeds into mathematics itself to change what it is in the future. The fields of mathematics and psychology do not describe pre-existing realities. Rather each field depicts realities that are consequential to *past* human endeavours or conceptualisations of what mathematics is and of what it is to be human. Mathematics is built to reflect the image we have of ourselves and becomes part of those selves that it reflects. The book concludes by suggesting that curriculum interventions, whether arising from new models of mathematics teacher education or from the influence of comparative testing, are not distortions of pre-existing conceptions of mathematics. Rather, they reflect new ways in which mathematics is evolving as a discipline, as a field of knowledge. Such interventions also produce revised conceptions of learners, teachers, teacher educators, researchers and of how policy works.

As outlined in the last two chapters, the theoretical centre of the proposed conclusion is in the work of Badiou. The Mathematics Education and Contemporary Theory conference, as an idea, had originally emerged out of a Badiou reading group comprising Julian, Yvette, our friend Rob Lapsley and myself. It was rare to find a book that so directly addressed both of our philosophical and mathematical interests.

> Badiou's canvas extends into the territory of potential futures, creating a framework against which all three twentieth-century philosophical traditions that he mentions can be read. His new book builds on his contention that these traditions were excessively centred on contemporary conceptions of the unit of the human, organised according to language-centred analyses. Badiou contends that truth is left out of this analytical mode. For Badiou scientific truth concerns the *invention* of theoretical parameters. "Truth can only be reached only through a process that breaks decisively with all established criteria for judging (or interpreting) the validity (or profundity) of opinions (or understandings) … access to truth can be achieved only by going against the grain of the world and against the current of history" (Hallward 2003, pp. xxiii-xxiv, see also pp. 209–221). Thus, truth cannot be substantiated or represented in culturally derived media. "Truths have no *substantial* existence" (Badiou 2009, p. 5, his emphasis). Any attempt to pinpoint truth ultimately disappoints us. So, in short, Badiou's quest is to understand how alternative forms of knowledge are shaped and evolve around a truth that is experienced but never finally represented (Brown, T. 2011, p. 149).

For Badiou (2009, p. 509), "History does not exist. There are only disparate presents whose radiance is measured by their power to unfold a past worthy of them". In my own work and in the last chapter, I have questioned to which extent are Piaget and Vygotsky are essential or overly prominent elements of mathematics education research (Brown, T. 2011). Or rather what are the consequences of assuming their prominence in shaping current debates. Is the insistence on their inclusion a normalising drag? Rather like in today's climate where it is impossible to avoid referencing comparative tests and thus seeing school mathematical learning as a race between nations to improve standards against a supposed model of improvement. Badiou asserts that there are multiple options in telling our history and evoking its potential elements to explain the current state of affairs. And in turn how do we tell stories about the present to justify our future actions? How can our disparate accounts be patched up into a new narrative to counter neoliberalism (Monbiot 2017) by introducing new histories that seek to include multiple priorities – not merely to identify the winners in prescribed forms of winning? Mathematics is often seen through the filter of mathematical experts and does not notice the potential that students can see. We can always map out the perceptual terrain differently. In psychoanalytic terms, we can always tell our past differently, to explain who we are now, rethink who we are now, so as to open new possible alternative future selves. We are *seemingly* subject to a history, but *who* created that history? And for what purpose? History has displayed a persistent and, dare I say, pernicious capacity to normalise the unacceptable. History, *if* it is supposed to exist, has not always served us very well. And there is some chance that it may not serve our futures very well. Might we revisit our storytelling potentials to open fresh ways of being? For Badiou, history needs a world for it to exist. History is a function of the world as we presently understand it. We have some capacity to decide how we understand that world and to make it happen. The history of mathematics education research has some very familiar points of anchorage. Those points of anchorage have marked out the field, but in some ways, they have also deadened the field. They have defined what is important but also policed what is important. History has entrapped us into oppressive ways of seeing. We must free our minds so that this entrapment does not exceed its purpose. Can theory assist us as we adjust to new conditions and to new conceptions of mathematics education research? Psychoanalytically speaking, can we rewrite the alleged past to open new futures?

9.1 The Becoming of Mathematics[1]

I have argued that the specific administration of school mathematics and associated teacher education described above changed conceptions of mathematics in our locale. From a wider perspective, we see the conceptions of school mathematics that influence our actions as a function of the discursive environment and the way that that environment formats mathematical activity. That is, the way in which mathematics is administered and conceptualised in the specific pedagogical environment determines what mathematics is. In England, successive governments have each followed rather authoritarian modes of curriculum definition and teacher education in the name of collective success. This approach feeds into public and individual perceptions of what mathematics is. We are part and parcel of a population where the majority of people understand the scope and purpose of mathematics through the filter of their own school education. Moreover, our sense of who we are is built through our own practice and the linguistic categories available to us, which are conditioned by particular, culturally preferred ways of making sense.

[1] This section draws on material first published as: Brown, T., Hodson, E., & Smith, K. (2013). TIMSS mathematics has changed real mathematics forever. *For the Learning of Mathematics, 33*(2), 38–43.

Formulations of pedagogical objects in mathematics such as the reconfiguring of mathematical tasks to meet new curriculum demands comprise qualitative adjustments to mathematical objects more generally. Changes to school mathematics can impact on mathematics itself. "Mathematics", of course, requires some qualification, to ensure that it has any meaning at all. Mathematics could mean the vast field of mathematics that is beyond the scope of any one individual. Alternatively, mathematics could be the majority view of what mathematics is and how it manifests itself in everyday practices across given populations. There cannot be final agreement on this. A choice needs to be made as to what counts as mathematics in any situation, and this choice always depends on circumstances that transcend most conceptions of mathematics. Nevertheless, it seems clear that the field of mathematics itself *has* been transformed through certain areas being explored more than others, for example, recently in government funding, statistics rather than geometry. Of course, most people do not explore many of these areas and so enjoy a relatively restricted view of mathematics.

School mathematics is susceptible to regular makeovers based on curriculum changes that redefine its content and preferred points of reference. New mathematical priorities, such as the need to meet the demands of international comparative testing, have come into prominence, whilst others have faded, such as problem-solving-based activity. A casual inspection of school textbooks or exam papers through successive decades would evidence a substantial shift. Mathematical objects are converted into pedagogical objects or standardised test items, which then influence the form of school mathematics (Morgan et al. 2011). School testing regimes have increasingly partitioned mathematical activity so that children and teachers are better able to successfully meet the priorities of international comparison (Askew et al. 2010; Brown, T. 2011). Further, key examinations for 16-year-olds have been shaped to meet new demands but have reduced students' capabilities and dispositions towards further study in mathematics (Pampaka et al. 2012). Test scores, enjoyment and functionality can pull in different directions.

Teacher education models meanwhile have been modified to secure greater compliance with those curriculum demands. And these new models of teacher education practice impact on wider understandings of mathematics, mathematics teaching, mathematics teacher education and mathematics education research. In describing changes to teacher education and its impact on conceptions of mathematics, our opinions are inevitably referenced to our own experience. We are responsible for overseeing the mathematics teacher education of a set of students. In my local world, teachers get to be teachers by following the route that has been briefly described in Chap. 5. The university impact on the experience of the student teachers is limited, as a consequence of spending so little time with them. Whatever our own mathematical credentials, or fantasies of what might be achieved in other circumstances, our everyday challenge is to attend to what might be achievable within the model that governs our practice and to work according to the outcomes of that model. Insofar as we are part of the community leading attempts to improve mathematics education, we may adopt a critical or resistant attitude in our efforts, but we must face up to the net effect of our actions, even though so many decisions have been taken out of our hands. Student teachers, meanwhile, develop a specific, pedagogically oriented conception of mathematics. They conceptualise their teaching within a rather restricted model of education. Blue skies are hidden by clouds.

To represent mathematics as universal, spanning nations and generations, comes at a price. Yet the resultant conceptions of school mathematics now define and regulate the boundaries of school mathematics. School mathematical knowledge derives from this newly described world backed up by governments using these conceptions of mathematics to set their policies and to materialise these new understandings of mathematics. These policy priorities may often exclude some powerful and interesting areas of mathematics. The revised priorities may then grow up to police the practices now developed in the name of mathematics (Kanes et al. 2010). Teachers are subject to skills criteria referenced to the curriculum success of their pupils. Pupils are understood through the grades they receive. The social parameters that govern our actions move on, as do the mathematical activities that are subject to these parameters.

Mathematics evolves through successive attempts to capture its objects (as defined by Badiou in an earlier chapter), such as through reaching new generalisations in newly encountered conditions. The advance of mathematics is defined by the production of such objects, often in response to newly defined objectives or pedagogical circumstances. Certain elements of mathematics have been touched more frequently by pedagogical or practical concerns. The field of mathematics has been marked out according to how it supports practical agendas. Some bits (e.g. statistics) are much more popular than other bits (e.g. topology) and for this reason tend to be more likely to be noticed in schools, used in everyday life, secure research grants, etc. Yet, it is actually quite difficult to sort mathematics according to which bits are empirically referenced like circles and which bits are not so common in appearance or utility, such as stellated octahedra. The historical circumstances that generated mathematical objects may have become a part of who we are such that we are no longer able to see them. Mathematical models exist as knowledge that sometimes supports empirical enterprises to certain limits, but, ultimately, as empirical support, the models always reach their limits.

Mathematics is reshaped to meet pedagogical or practical demands. Is mathematics then merely a social construction linked to our practical ambitions that anchor the existence of mathematics and the existence of its objects? Or does it have some underlying truth that, as it were, holds mathematics in place? This question requires a deeper philosophical analysis of whether there is mathematics beyond all of our socially motivated encapsulations. Similarly, humans are social constructions as a consequence of particular attributes being privileged in our understanding of them. Social constructions can change what humans are.

Both mathematics and psychology describe realities that are consequential to past human endeavours or conceptualisations. As fields of enquiry, they are enterprises, or objects, built *in* the human's own self-image (and then built *into* the human self-image) that trap us into thinking that there are universal realities of what it is to be human and of what it is to be mathematical. TIMSS and PISA, for example, are in the business of serving an image of mathematics characterised by particular forms of questioning, and, by serving that image, they make mathematics itself seem more real or part of a more enduring reality.

The learning and teaching of mathematics may be helpfully understood as seeing and experiencing mathematics as coming into being or participating in the becoming of mathematics, making it come into being (e.g. Krummheuer 2009; Roth 2010a). The learner may experience mathematics as part of herself, a self that is also evolving in the process. Mathematical objects and the ways in which we relate to them would never finally settle in relation to each other. The building of mathematics then reflects the image we have of ourselves and becomes part of those selves that it reflects. I have considered how conceptions of school mathematics change as a result of rethinking the needs of teacher knowledge. I am now taking a broader historical perspective. Yet, we may not experience our immersion in mathematical changes in this way. We understand ourselves as operating in a rather more restrictive space decided upon by legislation and by expectations beyond our active control. If the world is built in our own image, our children may encounter that world as an external demand out of line with their own perceived needs. Following Hegel, Malabou (2011) suggests that the individual "does not recognise itself in the community that it is nevertheless supposed to have wanted [...] The individual is 'alienated from itself'" (p. 24). The "self is already implicated in a social temporality that exceeds its own capacities for narration" (p. 28). These fractures in our self-image can result in adjustments to our tangible reality and to how we encounter it. Mathematics is a function of how we organise its supposed content at any point in time. Yet it is also a function of the narratives that report on how we experience it through time and of the hermeneutic working of those narratives that generate new dimensions of mathematics (Doxiadis and Mazur 2012). These narratives may be productive, misguided, manipulative or functions of particular administrative perspectives. For example, ideal accounts of mathematics can readily become policing structures in the service of compliant behaviour transforming how subsequent students experience mathematics. Curriculum innovation and associated testing can activate new, perhaps unexpected, modes of mathematical engagement or educative encounters across a community. People or communities more or less identify with these new conceptions of mathematics and shape their practices accordingly. For example, the model of teacher education described earlier reduces options for student teachers to see beyond compliance with the current curriculum and the associated assessment that seeks to mirror those international ambitions. Similarly, TIMSS and PISA have shaped mathematics through their widespread influence over how mathematics is understood, how it is conducted, how it is reproduced, thus rewriting what it is. As a result they have influenced the demands placed on teachers and students thus shaping who they are, perhaps locking them into unhelpful caricatures that can hinder adjustment to new circumstances. Meanwhile, mathematics education researchers can be cast in terms of supporting this kind of agenda. Many grant applications, for example, are oriented towards improving performance in a given regime, steering school learning back to a correct path more in tune with "what mathematics really is". The challenge for researchers in mathematics education is to recognise their political role in understanding how changing circumstances shape both mathematics and humans.

For Lacan (2008) knowledge is always renewable. But that knowledge provides much of the everyday reality in which we believe. We can't cope without that every day security of a story that usually works or seems to. The emphasis on the stories that we tell is not to suggest that we reject the knowledge that we have. That is our reality after all. We can learn from how those imposed stabilities guide life, or sometimes, whole lives. We may assume particular discursive formations, familiar names or labels, set rules, introduce analytical frameworks or hold certain assumptions for the time being, which influence the questions that we ask. For such knowledge is a function of the *worlds* in which we live. Indeed, much of our infrastructure (buildings, modes of governance, law, social practices, preferred styles, pedagogical objects, curriculum forms, schools, conceptions of teacher, examinations) is a function or reification of how previous generations conceptualised life. Knowledge is built into to our physical environment. It is also built into us as inhabitants of that space. We can however better appreciate the limits and limitations of such worlds and the forms of knowledge that they host to avoid the false comfort in contingent arrangements and to better understand how those arrangements shape or restrict our actions. The story or image never lasts. It always needs to be renewed. Learning might be understood as being about constant adjustment to a new mode of apprehension. For Lacan (2008, p. 17), "Truth is always new, and for it to be true it has to be new" because life as lived always exceeds the models that we try to place upon it. And the failures of these models as we use them produce desire to get things right. Lacan's subject of desire is always reaching beyond the current state of affairs, a perpetual quest to improve on the current story motivated by spotting the "holes in discourse" (Lacan 2008, p. 27). A learner would then be seeing and experiencing the world as coming into being, experiencing aspects of this world as part of herself, a self that is also evolving in the process.

Except not! So often past versions of knowledge are enshrined for well beyond their sell by date. We get stuck with them and they can underpin reactive tendencies. Lacan's conception of knowledge renewal entails detecting the limits and limitations of the ideological parameters that shape our actions and function as resistance to change. His work is centred on the potentialities of noticing holes in our current story and how these blind spots might alert us to new perspectives, to new ways of being. Any world relates to a state of knowledge. Knowledge, however, does not capture truth (for all), and for this reason knowledge will always need to be revised to better fit the times and circumstances. But actually we don't necessarily want a good fit. Rather the inevitable gap between story and what it seeks to portray is productive of desire, which might be seen as life itself. Having got any story right implies that the death drive has succeeded, or reactionary forms of governance have become entrenched. Pais (2015, p.) argues:

> a radical use of social theory in mathematics education gains from conceiving the importance of mathematics not in terms of mathematics itself, but in terms of the place this subject occupies within a given structural arrangement. There is something inherently wrong in the way researchers use social theory, and still behave as ambassadors of mathematics. No matter how much we would like mathematics to be an adventure into knowledge, the ultimate problem-solving technology or a crucial dimension of critical citizenship, this is not what school mathematics is. I suggest that school mathematics should be investigated as a crucial element of today's political and economic landscape, and not so much, as it is today, as a precious knowledge aimed to empower people and to enable societal development.

As mentioned, for Pais, cultural renewal or societal developments are ideological notions that can shield the real operation of politics.

The narratives that we offer about who we are never catch up with us, but that does not stop us from trying. And our misses can still be informative about who we are, when we are or where we are or where we are stuck. Narratives shape our desires, even if they do not take us to the place that satisfies them. The stories we tell do not pin down life for inspection but rather can stimulate this life for future growth if we do not get locked into old habits. There is always a risk that we begin to believe the stories we tell; they produce and control life rather than report on it, as though they provide the final answer (such as that the school curriculum effectively encapsulates mathematics, that sufficient research evidence would enable more certain action, that if I responded positively to all the demands placed upon me that I would be a good teacher, content with life, etc.). There is a cost for the individual as a result of gearing into the shared outer world. Through expressing oneself through social codes and procedures, personal and social boundaries are reshaped causing a troubling compulsion to settle these boundaries. The subject, or the "human", or the "personality", in Lacanian terms derives from persistent (failed) attempts to make sense of the world. This process is akin to the dialectic often associated with Hegel. We can never get our story to feel quite right. For each new assertion produced in this process, the outside, or the negation, of that assertion is also transformed:

> It is a new concept but one higher and richer than the preceding – richer because it contains or opposes the preceding and therefore contains it, and it contains even more than that, for it is the unity of itself and its opposite. (Hegel 2010, p. 33).

The "negative" is not imposed upon the object from the outside but is instead inherent to the concept itself. As Pais (in conversation) put it: "A totality, in order to truly be a totality, has also to include its own "exception", its negation. … Hegel's dialectics is about something being two "things" at the same time (itself and its negation)". "Something is a transition"(Hegel 2010, p. 90). "Hegelian dialectics is a kind of hysterical undermining of the master.., the immanent self-destruction and self-over-coming of every metaphysical claim. In short, Hegel's 'system' is nothing but a systematic tour through the failures of philosophical projects". (Žižek 2017, p. 4).

School mathematics is built in the human's own self-image as it reflects the practical challenges it is supposed to serve. Humans, however, are a product of the worlds that they have produced; unequal, troubled, more or less functional and normal. The mathematical concepts that they have constructed to meet the challenges are built *into* the human self-image, of who they are or who they may want to be. These parallactic self-producing and self-validating rationalities trap us into believing that there are universal realities (or rationalities) as to what it is to be mathematical and as to what it is to be human that become so many targets or aspirations and thereby deflect us from our everyday political entrapments. Rationalities are then produced that are particular to those contingent arrangements or understandings of the world. The material points of reference that characterise school mathematics then support both a speculative belief in mathematical concepts referenced to contemporary societal structures and a contingent rationality that connects them.

→ None of this is new to a Foucault scholar!!

Seeing the project of mathematics education as so many classrooms organising the activity of individual children is only one version of understanding how mathematical learning is taking place in the world today. Individuals respond to the demands that they encounter, but does the master (the educational system, capitalism, democracy) really know what he wants? We are taught to desire through our cultural engagements, but desire mistakes its object, setting its targets in slightly unpredictable ways. We aspire to futures that would never quite satisfy us if we actually got there. It was this ever desiring, never satisfied, being derived from Freud that fed into early conceptions of people living within twentieth-century capitalist economics (Garetsky 2004). Mathematics is desired by society. This desire to have mathematics gets expressed as a demand for something more specific, such as a set of particular skills, or a curriculum of a certain form. We might ask, however, why the desires expressed *in* a demand for certain sorts of school mathematics get expressed *as* a demand for certain sorts of school mathematics. In many contemporary contexts, mathematics has come to be defined as *an end result* of an intellectual process rather than as *the process* of getting there. School curricula now emphasise skills rather than deeper appreciation, "doing" rather than "interpreting". University mathematics has so often become a survey of classic end points rather than a regeneration of mathematical experience. The symbolisation teaches us what we already know but introduces reification and thus a commodification and in due course mummification. This packaging of mathematical activity is a relatively arbitrary function of the supposed fulfilled fantasy of what mathematics really is.

There are many instances where the student's demonstration of his or her mathematical understanding amounts to little more than filling in a gap in a story provided by some sort of assessment device. Countless studies in mathematics education research have sought to isolate the mathematical dimension of wider discussion. But such questions are crucially linked to the geography of the supposed interface of human subject and object. Mathematics in schools exists substantially as pedagogical material crafted for supposed modes of apprehension or consumption. Students are required to spot certain things according to the given mark scheme. But such apprehension depends on how we understand mathematical objects and how we understand human subjects. If you fix one, then you fix the other. That is, a given mark scheme supposes a given conception of a student able to answer on those terms and supposes that mathematics can be seen in a definitive way and taught by a teacher able to evoke it in that way. Framed in this way, a mathematical idea, as encountered in school, is little more than a placeholder for externally motivated desire producing somewhat arbitrary demands. We can know mathematical ideas in the sense that our knowledge can be verifiable against familiar approaches. But the familiar can trap us into reactive ways of making sense.

In line with the take-away message promised earlier, the production of mathematical concepts may be helpfully understood as seeing and experiencing mathematics as coming into being or participating in the becoming of mathematics, making it come into being. The learner may experience mathematics as part of herself, a self that is also evolving in an essentially psychoanalytic process where conceptions of the social are persistently renewed. Mathematical concepts and the ways in which we relate to them would never finally settle in relation to each other nor do the theories we introduce. Their final form always stays out of reach. The building of mathematics then reflects the image we have of ourselves and becomes part of those selves that it reflects.

School mathematics does not generally reach for the stars and often prefers to make do with some rather rusty scaffolding in the name of corrosive metrics. There is a recurrent sense that there should have been more to it than has been allowed. Whilst the truth of mathematics can sometimes be used to underwrite its ideologically motivated manifestations, we need to trouble the "truths" that are presented to us, towards encountering the spaces beyond and the hold they have on us.

GOREN FARM SEEDS

Wild Flowers

Premium Mix

54 species of Native flowers and grasses
For habitat restoration, suitable for a
wide range of acid alkaline soil types.

Premium Wild Flower Meadow Mix

Contains 38 species of native flowers : 5% Yellow Rattle, Knapweed, Cat's-ear, Rough & Autumn Hawkbit, smooth Hawks Beard, Ribwort Plantain, Wild Red & White Clover, Red Sorrel, Birdsfoot Trefoil, Common Spotted Orchid, Field Forget-me-not, common & bush vetch, Meadow & Grass Vetchling, Black Medic, Selfheal, Betony, Bulbous and Meadow Buttercup, Cowslip, Wild Carrot, Yarrow, Ox-Eye Daisy, Goats Beard , Goats Beard. Common Agrimony, Pignut, Evening primrose, Musk Mallow, Red Campion, Cow Parsley, Devils Bit, Ragged Robin, Meadow Cranesbill. 6 **Cornfield annuals**: Cockle, Poppy, Flower, Camomile, Marigold ,wild pansy+ 10 **grasses**: Smooth stalked meadow , Sweet vernal, Quaking , Crested dogtail, Sheeps , hard, Slender red fescue,s Meadow barley, Yellow oatgrass, Rough bent Grass .

1. Scarify lightly, remove moss and thatch, spread seed on surface and in roll very well. Some of the varieties sown in the spring will need to over winter to germinate the following spring.

2. Cut (and Remove if possible) or graze any Grass re-growth in Dec/Jan to promote seed germination from February. Remove any loose leaves where trees are nearby

3. Plants should be allowed to set seed from early July to propagate itself for the following year. Knock the seed out.

4. Annually, Grass can be cut and removed from mid July-October and then regrowth cut or grazed up till January. Sometimes cut in June to promote later flowering plants.

5. Sowing Rate: at **5g- 10g/m² & flowers only = 0.5-1g/m²**

Origin :**Goren Farm** , Stockland, Honiton, Devon, EX 149 EN
www.goren.co.uk

References

Alcorn, M. (1994). The subject of discourse: Reading Lacan through (and beyond) poststructuralist contexts. In M. Bracher, M. Alcorn, R. Corthell, & F. Massardier-Kenney (Eds.), *Lacanian theory of discourse: Subject, structure, and society* (pp. 19–45). New York: New York University Press.

Alexander, R. (Ed.). (1990). *Children, their world, their education: Final report and recommendations of the Cambridge Primary Review*. London: Routledge.

Althusser, L. (1971). Ideology and ideological state apparatuses. In *Lenin and philosophy and other essays*. Translated and edited by B. Brewster. London: New Left Books. (pp. 232–272).

Althusser, L. (2014). *On the reproduction of capitalism: Ideology and ideological state apparatuses*. (G. M. Goshgarian Trans). London: Verso.

Andrade-Molina, M., Valero, P., & Ravn, O. (2019). The amalgam of faith and reason: Euclid's elements and the scientific thinker. In P. Ernest (Ed.), *The philosophy of mathematics education today* (pp. 99–113). New York: Springer.

Askew, M. (2008). Mathematical discipline knowledge requirements for prospective primary teachers, and the structure and teaching approaches of programmes designed to develop that knowledge. In P. Sullivan & T. Woods (Eds.), *Knowledge and beliefs in mathematics teaching and teaching development. The international handbook of mathematics teacher education* (Vol. Volume 1, pp. 13–35). Rotterdam: Sense Publishers.

Askew, M., & Venkat, H. (2019). Mathematical subject knowledge for teaching primary school mathematics. In D. Potari & O. Chapman (Eds.), *The international handbook of mathematics teacher education* (pp. 15–41). Rotterdam: Brill.

Askew, M., Hodgen, J., Hossain, S., & Bretscher, N. (2010). *Values and variables: A review of mathematics education in high-performing countries*. London: The Nuffield Foundation.

Atkinson, D. (2018). *Art, disobedience and ethics: The adventure of pedagogy*. Cham, Switzerland: Palgrave Macmillan.

Badiou, A. (2007). *Being and event*. London: Continuum.

Badiou, A. (2009). *Logics of worlds*. London: Continuum.

Badiou, A. (2011). *Second manifesto for philosophy*. Cambridge: Polity Press.

Ball, D. L., Hill, H. C., & Bass, H. (2005). Knowing mathematics for teaching: Who knows mathematics well enough to teach third grade, and how can we decide? *American Educator, Fall*, 14–17. 20–22, 43–46.

Ballantyne, M., & Brown, T. (2016). Close encounters of the curved kind. *Mathematics Teaching*. September, 28–30.

Barad, K. (2007). *Meeting the universe halfway: Quantum physics and the entanglement of matter and meaning*. Durham, NC: Duke University Press.

T. Brown, *A Contemporary Theory of Mathematics Education Research*,
https://doi.org/10.1007/978-3-030-55100-1

Barthes, R. (1977). *Image, music, text*. Glasgow: Fontana.

Barton, B. (2008). *The language of mathematics: Telling mathematical tales*. Dordrecht: Springer.

Barwell, R. (2019). Some thoughts on a mathematics education for environmental sustainability. In P. Ernest (Ed.), *The philosophy of mathematics education today* (pp. 145–160). New York: Springer.

Bastani, A. (2019). *Fully automated luxury communism: A manifesto*. London: Verso.

Bauman, Z. (2014). Lecture at *Politics in times of anxiety conference*, Manchester University, 9th June.

Bhabha, H. (1994). *The location of culture*. London: Routledge.

Bibby, T. (2010). *Education – An 'impossible profession'?: Psychoanalytic explorations of learning and classrooms*. London: Routledge.

Bikner-Asbahs, A., Vohns, A., Bruder, R., Schmitt, O., & Dorfler, W. (2016). *Theories in and of mathematics education: Theory strands in German speaking countries*. Springer Cham Switzerland.

Black, L., Mendick, H., & Solomon, Y. (Eds.). (2009). *Mathematical relationships: Identities and participation*. London: Routledge.

Bloom, B. (1968). Learning for mastery. *Evaluation Comment, 1*(2) 1–12.

Boylan, M., & Adams, G. (2019). Paradox, market mirages and the stratification of professional learning: the case of mathematics education professional development in England. Paper presented at: *BERA conference 2019*. Manchester University, Manchester, September 2019.

Boylan, M., Maxwell, B., Wolstenholme, C., Jay, T., & Demack, S. (2018). The mathematics teacher exchange and 'mastery' in England: The evidence for the efficacy of component practices. *Education Sciences, 8*(4), 202.

Bracher, M. (1994). On the psychological and social functions of language: Lacan's theory of the four discourses. In M. Bracher, M. Alcorn, R. Corthell, & F. Massardier-Kenney (Eds.), *Lacanian theory of discourse: Subject, structure, and society* (pp. 107–128). New York: New York University Press.

Brown, M. (2011). Going back or going forward? Tensions in the formulation of a new national curriculum in mathematics. *Curriculum Journal, 22*(2), 151–165.

Brown, T. (1987a). Teacher student interactions. *For the Learning of Mathematics, 7*(1), 48–52.

Brown, T. (1987b) *Language patterns in lessons featuring mathematical investigations*. PhD thesis. Southampton University.

Brown, T. (1997). *Mathematics education and language: Interpreting hermeneutics and post-structuralism*. Dordrecht: Springer.

Brown, T. (2001). *Mathematics education and language: Interpreting hermeneutics and post-structuralism. Revised second edition*. Dordrecht: Springer.

Brown, T. (Ed.). (2008a). *The psychology of mathematics education: A psychoanalytic displacement*. Rotterdam: Sense Publishers.

Brown, T. (2008b). Signifying "learner", "teacher" and "mathematics": A response to a special issue. *Educational Studies in Mathematics, 69*, 249–263.

Brown, T. (2008c). Lacan, subjectivity and the task of mathematics education research. *Educational Studies in Mathematics, 68*, 227–245.

Brown, T. (2008d). Desire and drive in researcher subjectivity: The broken mirror of Lacan. *Qualitative Inquiry, 14*(3), 402–423.

Brown, T. (2011). *Mathematics education and subjectivity: Cultures and cultural renewal*. Dordrecht: Springer.

Brown, T. (2012). Subjectivity and cultural adjustment in mathematics education: A response to Wolf-Michael Roth. *Educational Studies in Mathematics, 80*(3), 475–490.

Brown, T. (2016). Rationality and belief in learning mathematics. *Educational Studies in Mathematics, 92*(1), 75–90.

Brown, T. (2018). *Teacher education in England: A critical interrogation of school-led training*. London: Routledge.

Brown, T., & Clarke, D. (2013). Institutional contexts for research in mathematics education. In M. A. Clements, A. Bishop, C. Keitel, J. Kilpatrick, & F. Leung (Eds.), *Third international handbook of mathematics education* (pp. 459–484). Dordrecht: Springer.

Brown, L., & Coles, A. (2012). Developing 'deliberate analysis' for learning mathematics and for mathematics teacher education: How the enactive approach to cognition frames reflection. *Educational Studies in Mathematics, 80*(1–2), 217–231.

Brown, T., & Jones, L. (2001). *Action research and postmodernism: Congruence and critique*. Buckingham: Open University Press.

Brown, T., & McNamara, O. (2011). *Becoming a mathematics teacher: Identity and identifications*. Dordrecht: Springer.

Brown, T., & Walshaw, M. (Eds.). (2012). Special issue: Mathematics education and contemporary theory. *Educational Studies in Mathematics*, *80*(1–2).

Brown, T., Atkinson, D., & England, J. (2006). *Regulatory discourses in education: A Lacanian perspective*. Bern: Peter Lang publishers.

Brown, T., Heywood, D., Solomon, Y., & Zagorianakos, A. (2013a). Experiencing the space we share. *ZDM: The International Journal of Mathematics Education, 45*, 561–572.

Brown, T., Hodson, E., & Smith, K. (2013b). TIMSS mathematics has changed real mathematics forever. *For the Learning of Mathematics, 33*(2), 38–43.

Brown, T., Rowley, H., & Smith, K. (2014). Rethinking research in teacher education. *British Journal of Educational Studies, 62*(3), 281–296.

Brown, T., Rowley, H., & Smith, K. (2015). Sliding subject positions: Knowledge and teacher educators. *British Educational Research Journal, 42*(3), 492–507.

Brown, T., Solomon, Y., & Williams, J. (2016). Theory in and for mathematics education: In pursuit of a critical agenda. *Educational Studies in Mathematics, 92*(3), 287–297.

Brown, T., Dore, M., & Hanley, C. (2019). *Research on becoming an English teacher: Through Lacan's looking glass*. London: Routledge.

Brown, T., Solomon, Y., & Williams, J. (Eds.). (2012) Special issue: Mathematics education and contemporary theory 2. *Educational Studies in Mathematics, 92*(3).

Bruner, J. S. (1966). *Toward a theory of instruction*. Cambridge, MA: Harvard University Press.

Butler, J. (1997). *The psychic life of power*. Stanford: Stanford University Press.

Butler, J. (2005). *Giving an account of oneself*. New York: Fordham University Press.

Cabral, T., & Baldino, R. (2019). The credit system and the summative assessment splitting moment. *Educational Studies in Mathematics, 102*(2), 275–288.

Calder, N. (2012). The layering of mathematical interpretations through digital media. *Educational Studies in Mathematics, 80*(1–2), 269–285.

Calder, N., & Brown, T. (2010). Learning through digital technologies. In M. Walshaw (Ed.), *Unpacking pedagogy: New perspectives for mathematics classrooms* (pp. 223–244). Charlotte, NC: Information Age Publishing.

Chap Sam, L. (2012). Public images of mathematics. *Philosophy of Mathematics Education Journal, 15*.

Chapman, O. (2019). Narratives in mathematics teacher education. In S. Llinares & O. Chapman (Eds.), *International handbook of mathematics education* (Vol. Volume 2, pp. 13–38). Rotterdam: Brill.

Charalambous, C., & Delaney, S. (2019). Mathematics teaching practices and practice based pedagogies. Review of the literature since 2000. In D. Potari & O. Chapman (Eds.), *The international handbook of mathematics teacher education*. Rotterdam: Brill.

Clandinin, D. J., & Connelly, F. M. (1992). Teacher as curriculum maker. In P. W. Jackson (Ed.), *Handbook of research on curriculum* (pp. 363–401). New York: Macmillan.

Clandinin, D. J., Long, J. S., Schaefer, L., Downey, C. A., Steeves, P., Pinnegar, E., McKenzie Robblee, S., & Wnuk, S. (2015). Early career teacher attrition: Intentions of teachers beginning. *Teaching Education, 26*(1), 1–16.

Clarke, M. (2012). The other side of education: A Lacanian critique of neoliberal education policy. *Other Education: The Journal of Educational Alternatives, 1*(1), 46–60.

Cobb, P., & Bowers, J. S. (1999). Cognitive and situated learning perspectives in theory and practice. *Educational Researcher, 28*(2), 4–15.

Collingwood, R. G. (1982). *An autobiography*. Oxford: Clarendon Press.

Confrey, J. (1991). Steering a course between Vygotsky and Piaget. *Educational Researcher, 20*, 28–32.

Cooper, B., & Dunne, M. (1999). *Assessing children's mathematical ability*. Buckingham: Open University Press.

Corfield, D. (2012). Narrative and the rationality of mathematical practice. In A. Doxiadis & B. Mazur (Eds.), *Circles disturbed: The interplay of mathematics and narrative* (pp. 244–280). Princeton: Princeton University Press.

Critchley, S. (2008). *Infinitely demanding: Ethics of commitment, politics of resistance*. London: Verso.

Davies, B. (2006). Subjectification: The relevance of Butler's analysis for education. *British Journal of Sociology of Education., 27*(4), 425–438.

Davis, C. (2009). Introduction: Holy Saturday or resurrection Sunday? Staging an unlikely alliance. In S. Žižek & J. Millbank (Eds.), *The monstrosity of Christ: Paradox or dialectic?*. (Ed. C. Davis) (pp. 2–23). Cambridge, MA: The MIT Press.

Davis, B., & Simmt, E. (2006). Mathematics-for-teaching: An ongoing investigation of the mathematics that teachers (need to) know. *Educational Studies in Mathematics, 61*, 293–319.

Dawkins, R. (2006). *The God delusion*. London: Bantam.

de Freitas, E. (2013). The mathematical event: Mapping the axiomatic and the problematic in school mathematics. *Studies in Philosophy and Education, 32*, 581–599.

de Freitas, E., & Nolan, K. (Eds.). (2008). *Opening the research text: Critical insights and in(ter) ventions into mathematics education*. New York: Springer.

de Freitas, E., & Sinclair, N. (2012). Diagram, gesture, agency: theorizing embodiment in the mathematics classroom. *Educational Studies in Mathematics, 80*(1–2), 133–152.

de Freitas, E., & Sinclair, N. (2014). *Mathematics and the body: Material entanglements in the classroom*. New York: Cambridge University Press.

de Freitas, E., & Walshaw, M. (2016). *Alternative theoretical frameworks of mathematics education research: Theory meets data*. Switzerland: Springer.

de Freitas, E., Sinclair, N., & Coles, A. (Eds.). (2017). *What is a mathematical concept?* New York: Cambridge University Press.

Deleuze, G., & Guattari, F. (1996). *What is philosophy?* London: Verso.

Department for Education. (2010). *The importance of teaching*. London: The Stationary Office.

Department for Education. (2012). Call for "Expressions of interest" for: Teaching Schools evaluation.

Department for Education. (2016). *South Asian method of teaching maths to be rolled out in school. Press release*. [Online] [Accessed on 9th October 2018] https://www.gov.uk/government/news/south-asian-method-of-teaching-maths-to-be-rolled-out-in-schools

Derrida, J. (1978). *Writing and difference*. Chicago: The University of Chicago Press.

Derrida, J. (1994). *The Specters of Marx: The state of the debt, the work of mourning, and the New International*. London: Routledge.

Dimotropoulis, A; ENTEP (Online). The Bologna Process and teacher education structures in Europe: Creating a European teacher education area. *European Network on Teacher Education Policies* (accessed 2016).

Dowling, P. (1998). *The sociology of mathematics education*. London: Falmer.

Doxiadis, A., & Mazur, B. (Eds.). (2012). *Circles disturbed: The interplay of mathematics and narrative*. Princeton: Princeton University Press.

EEF (Education Endowment Fund). (2017). *Improving mathematics in key stages two and three: Guidance report*. Education Endowment Fund and Sutton Trust.

Ellis, V., Glackin, M., Heighes, D., Norman, M., Norris, K., Spencer, I., & McNicholl, J. (2013). A difficult realisation: The proletarianisation of higher education-based teacher educators. *Journal of Education for Teaching: International Research and Pedagogy, 39*(3), 266–280.

Emerson, C. (1983). The outer world and inner speech: Bakhtin, Vygotsky, and the internalization of language. *Critical Inquiry, 10*(2), 245–264.

ENTEP: Dimitropoulos, A. (accessed 2020). The Bologna process and teacher education structures in Europe: Creating a European teacher education area . European Network on Teacher Education Policies. http://entep.unibuc.ro/wp-content/uploads/2017/07/A.-Dimitropoulos-2008.-The-Bologna-process-and-teacher-education-structures-in-Europe-Creating-a-European-Teacher-Education-Area.pdf.

Ernest, P. (2016). The problem of certainty in mathematics. *Educational Studies in Mathematics, 9*(3), 379–393.

Ernest, P. (2019a). The ethics of mathematics: Is mathematics harmful? In P. Ernest (Ed.), *The philosophy of mathematics education today*. New York: Springer.

Ernest, P. (Ed.). (2019b). *The philosophy of mathematics education today*. New York: Springer.

Feltham, O. (2008). *Alain Badiou: Live theory*. London: Continuum.

Fielding, M., & Moss, P. (2011). *Radical education and the common school: A democratic alternative*. London: Routledge.

Fink, B. (1995). *The Lacanian subject: Between language and jouissance*. Princeton, NJ: Princeton University Press.

Fischbein, E. (1987). *Intuition in science and mathematics. An educational approach*. Dordrecht: Kluwer Academic Publishers.

Foucault, M. (2009). *Security, Territory, Population: Lectures at the College de France, 1977–78*. Basingstoke: Palgrave Macmillan.

Freud, S. (2002). *Civilisation and its discontents*. London: Penguin.

Garetsky, E. (2004). *Secrets of the soul: A social and cultural history of psycho-analysis*. Toronto: Knopf.

Garner, R. (2012). Almost half of adults have maths abilities of an 11-year-old, *The Independent* (3rd March).

Gibb, N. (2016). Building a renaissance in mathematics teaching. [Online] [Accessed on 22 September 2019] https://www.gov.uk/government/speeches/nick-gibb-building-a-renaissance-in-mathematics-teaching

Goldin, G., Rösken, B., & Törner, G. (2009). Beliefs – No longer a hidden variable in mathematical teaching and learning processes. In J. Maaß & W. Schlöglmann (Eds.), *Beliefs and attitudes in mathematics education: New research results* (pp. 1–18). Rotterdam: Sense.

Goos, M. (2005). A sociocultural analysis of the development of pre-service and beginning teachers' pedagogical identities as users of technology. *Journal of Mathematics Teacher Education, 8*(1), 35–59.

Griffiths, J. (2019). *The great firewall of China: How to build and control an alternative vision of the internet*. London: Zed.

Habermas, J. (1972). *Knowledge and human interests*. London: Heinemann.

Hallward, P. (2003). *Badiou: A subject to truth*. Minneapolis: University of Minnesota Press.

Hanley, C., & Brown, T. (2016). Shifting subject knowledge identifications for student teachers of English. *Journal of Curriculum Studies, 49*(3), 352–368.

Hanley, C., & Brown, T. (2017). Thinking with certainty or with doubt: A theorisation of discursive knowledge, with particular reference to the work of Jacques Lacan. *Discourse: Studies in the Cultural Politics of Education. 40*(3), 306–318.

Hannula, M. (2012). Exploring new dimensions of mathematics-related affect: Embodied and social theories. *Research in mathematics education, 14*(2), 137–161.

Harvey, D. (2015). *Seventeen contradictions and the end of capitalism*. London: Profile Books.

Hawking, S., & Mlodinow, L. (2010). *The grand design*. New York: Bantam.

Hegel, G. (2010). *Science of logic*. Cambridge: Cambridge University Press.

Heywood, D., & Parker, J. (2010). *The pedagogy of physical science*. New York: Springer.

Hill, H., Rowan, B., & Ball, D. (2005). Effects of teachers' mathematical knowledge for teaching on student achievement. *American Educational Research Journal, 42*(2), 371–406.

Hobson, A. J., Malderez, A., & Tracey, L. (2009). *Navigating initial teacher training: Becoming a teacher*. London: Routledge.

Hodson, E., Smith, K., & Brown, T. (2010). Reasserting theory in professionally-based initial teacher education. *Teachers and Teaching, 18*(2), 181–195.

Hough, S. (2012). *Fair shares: Fractions. Teacher book*. London: Hodder Education.
Hoyles, C., Noss, R., Kent, P., & Bakker, A. (2010). *Improving mathematics at work: The need for techno-mathematical literacies*. Abingdon: Routledge.
Hudson, B., & Zgaga, P. (Eds.). (2008). Teacher education policy in Europe: A voice of higher education institutions. *Monographs on Journal of Research in Teacher Education*, Faculty of Teacher Education, University of Umeå, Umeå, Sweden.
Jaworski, B. (2019). Inquiry based practice in university teaching practice development. In D. Potari & O. Chapman (Eds.), *The international handbook of mathematics teacher education* (pp. 335–361). Rotterdam: Brill.
Jones, M., & Straker, K. (2006). What informs mentors' practice when working with trainees and new teachers? An investigation in mentors' professional knowledge base. *Journal of Education for Teaching, 32*(2), 165–184.
Kanes, C., Morgan, C., & Tsatsaroni, A. (2010). 'Analysing PISA's regime of rationality', *Proceedings of the 6th Conference of Mathematics Education and Society*, Berlin.
Klein, M. (2012). How inconvenient assumptions affect preservice teachers' uptake of new interactional patterns in mathematics: Analysis and aspiration through a bifocal lens. *Educational Studies in Mathematics, 80*(1–2), 25–40.
Krummheuer, G. (2009). Inscription, narration and diagram-based argumentation: Narrative accounting practices in primary mathematics classes. In W.-M. Roth (Ed.), *Mathematical representation at the interface of body and culture*. Charlotte, NC: Information Age Publishing.
Kullberg, A., Runesson, U., & Marton, F. (2017). What is made possible to learn when using the variation theory of learning in teaching mathematics? *ZDM Mathematics Education, 49*(4), 559–569.
Lacan, J. (1986). *The four fundamental concepts of psycho-analysis*. London: Peregrine.
Lacan, J. (2000). *Psychoses: The seminar of Jacques Lacan. Book III*. London: Routledge.
Lacan, J. (2006). *Écrits*. (Bruce Fink Trans). New York: W. W. Norton.
Lacan, J. (2007). *The other side of psychoanalysis: The seminar of Jacques Lacan: Book XVII*. London: Norton.
Lacan, J. (2008). *My teaching*. London: Verso.
Laclau, E. (2007). *Emancipations*. London: Verso.
Lave, J., & Wenger, E. (1991). *Situated learning: Legitimate peripheral participation*. Cambridge: Cambridge University Press.
Leder, G. C., Pehkonen, E., & Törner, G. (2002). *Beliefs: A hidden variable in mathematics education?* Dordrecht: Kluwer Academic Publishers.
Lerman, S. (2000). A case of interpretations of 'social': A response to Steffe and Thompson. *Journal for Research in Mathematics Education, 31*(2), 210–227.
Lerman, S. (2020). *Encyclopedia of mathematics education*. Springer.
Llewellyn, A. (2018). *Manufacturing the mathematics child: A deconstruction of dominant spaces of production and governance*. London: Routledge.
Lundin, S. (2012). Hating school, loving mathematics: The ideological function of critique and reform in mathematics education. *Educational Studies in Mathematics, 80*(1–2), 73–85.
Malabou, C. (2011). Is confession the accomplishment of recognition? Rousseau and the unthought of religion. In S. Žižek, C. Crockett, & C. Davis (Eds.), *Hegel and the infinite: Religion, politics and dialectic*. New York: Columbia University Press.
Mason, J. (1989). Mathematical abstraction as the result of a delicate shift of attention. *For the Learning of Mathematics, 9*, 2–8.
Mason, J. (1994). Researching from the inside in mathematics education: Locating an I-you relationship, *Proceedings of the eighteenth conference of the group on the Psychology of Mathematics Education*, University of Lisbon, 1, 176–194.
Mathematics Mastery. (2020). *Secondary Classroom Resources*. [Online] [accessed on 22 May 2020] https://www.mathematicsmastery.org/classroom-resources-secondary-maths-teaching-support?phase=secondary&c=5ec7deb239305
McEwan, I. (2012). The originality of the species, *Guardian* (23 March).

McEwan, I. (2019). *Machines like me*. London: Jonathan Cape.
McLeod, D. (1992). Research on affect in mathematics education: A reconceptualisation. In D. Grouws (Ed.), *Handbook of research on mathematics teaching and learning*. New York: Kogan Page.
McNamara, D. (1991). Subject knowledge and its application: Problems and possibilities for teacher educators. *Journal of Education for Teaching, 17*(2), 113–128.
Monbiot, G. (2017). *Out of the wreckage: A new politics for an age of crisis*. London: Verso.
Morgan, C. (1998). *Writing mathematically*. London: Falmer.
Morgan, C., Tang, S., & Sfard, A. (2011). Grammatical structure and mathematical activity: Comparing examination questions. In C. Smith (Ed.), *Proceedings of the British Society for Research into learning mathematics*. 31(3), November.
Muis, K. R., Psaradellis, C., Lajoie, S. P., Di Leo, I., & Chevrier, M. (2015). The role of epistemic emotions in mathematics problem solving. *Contemporary Educational Psychology, 42*, 172–185.
NCETM. (2016). *The essence of maths teaching for mastery.* [Online] [Accessed on 11 November 2018] https://www.ncetm.org.uk/files/37086535/The+Essence+of+Maths+Teaching+for+Mastery+june+2016.pdf
NCETM. (2018). Teaching for Mastery: Supporting Research, Evidence and Argument. [Online] [Accessed on 12 November 2018] https://www.ncetm.org.uk/resources/50819#memorisation
NCETM. (2019a). *Five big ideas in teaching for mastery.* [Online] [accessed on 2 October 2019] https://www.ncetm.org.uk/resources/50042
NCETM. (2019b). *Supporting research, evidence and argument.* [Online] [Accessed on 1 November 2019] https://www.ncetm.org.uk/resources/50819
Nolan, K. (2016). Schooling novice mathematics teachers on structures and strategies: A Bourdieuian perspective on the role of 'others' in classroom practices. Special issue on contemporary theory II. *Educational Studies in Mathematics, 92*(3), 315–330.
Otte, M. (2019). The philosophy of mathematical education between Platonism and the computer. In Ernest, P. (Ed.). (2019b). *The philosophy of mathematics education today*. New York: Springer. (pp. 61–80).
Pahl, K. (2011). The way of despair. In S. Žižek, C. Crockett, & C. Davis (Eds.), *Hegel and the infinite: Religion, politics and dialectic* (pp. 141–158). New York: Columbia University Press.
Pais, A. (2013). An ideology critique of the use-value of mathematics. *Educational Studies in Mathematics, 84*, 15–34.
Pais, A. (2014). Economy: The absent Centre of mathematics education. *ZDM-The International Journal on Mathematics Education., 46*, 1085–1093.
Pais, A. (2015). Symbolising the real of mathematics education. *Educational Studies in Mathematics., 89*(3), 375–391.
Pais, A. (2016). At the intersection between the subject and the political: A contribution to an ongoing debate. *Educational Studies in Mathematics, 92*(3), 347–359.
Pais, A. (2019). Mathematics, capitalism and biosocial research. *Educational Studies in Mathematics, 101*(3), 373–386.
Pais, A., & Costa, M. (2018). An ideology critique of global citizenship education. *Critical Studies in Education* 1–16.
Pais, A., & Valero, P. (2012). Researching research: Mathematics education in the political. *Educational Studies in Mathematics, 80*(1–2), 9–24.
Palmer, A. (2011). "How many sums can I do"? Performative strategies and diffractive thinking as methodological tools for rethinking mathematical subjectivity. *Reconceptualizing Educational Research Methodology, 1*(1), 3–18.
Pampaka, M., Williams, J., Hutcheson, G., Wake, G., Black, L., Davis, P., & Hernandez- Martinez, P. (2012). The association between mathematics pedagogy and learners' dispositions for university study. *British Educational Research Journal, 38*(3), 473–496.
Paton, G. (2012). Numeracy campaign: 17m adults 'struggle with primary school maths', *The Telegraph.* (2 March).

Pawlik, P. (2020). *The discursive construction of the mastery curriculum in mathematics*. Doctor of Education thesis. Manchester Metropolitan University.
Piaget, J. (1952). *The child's conception of number*. London: Routledge.
Piketty, T. (2014). *Capital in the twenty-first century*. London: Harvard University Press.
Piketty, T. (2020). *Capital and ideology*. London: Harvard University Press.
Pimm, D. (1987). *Speaking mathematically*. London: Routledge.
Polanyi, M. (1961). *Personal knowledge: Towards a post-critical philosophy*. London: Routledge and Kegan Paul.
Potari, D., & Chapman, O. (Eds.). (2019). *The international handbook of mathematics teacher education*. Rotterdam: Brill.
Presmeg, N., & Radford, L. (2008). On semiotics and subjectivity: A response to Tony Brown's "signifying 'students', 'teachers', and 'mathematics': A reading of a special issue". *Educational Studies in Mathematics, 69*, 265–276.
Radford, L. (2016). On alienation in the mathematics classroom. *International Journal of Educational Research, 79*, 258–266.
Radford, L. (2018). Semiosis and subjectification: The classroom constitution of mathematical subjects. In N. Presmeg, L. Radford, M. Roth, & G. Kadunz (Eds.), *Signs of signification. Semiotics in mathematics education research*. Cham, Switzerland: Springer.
Radford, L. (2019). A plea for a critical transformative philosophy of mathematics. In P. Ernest (Ed.), *The philosophy of mathematics education today* (pp. 1–10). New York: Springer.
Ricoeur, P. (1981). *Hermeneutics and the human sciences*. Cambridge: Cambridge University Press.
Ricoeur, P. (1984). *Time and narrative* (Vol. 1). Chicago: Chicago University Press.
Roesken, B., Pepin, B., & Torner, G. (2011). Beliefs and beyond: Affect and the teaching and learning of mathematics. *ZDM: The International Journal on Mathematics Education, 43*(4), 451–455.
Roth, W.-M. (2010a). Thinking and speaking: A dynamic approach. In W.-M. Roth (Ed.), *Re/structuring science education: ReUniting sociological and psychological perspectives*. Dordrecht: Springer.
Roth, W.-M. (Ed.). (2010b). *Re/structuring science education: Reuniting sociological and psychological perspectives*. Dordrecht: Springer.
Roth, W.-M. (2012a). Societal mediation of mathematical cognition and learning. *Orbis Scholae, 6*(2), 7–22.
Roth, W.-M. (2012b). Re/writing the subject: A contribution to post-structuralist theory in mathematics education. *Educational Studies in Mathematics, 80*, 451–473.
Roth, W.-M., & Lee, Y. J. (2007). Vygotsky's neglected legacy: Cultural-historical activity theory. *Review of Educational Research, 77*, 186–232.
Roth, W.-M., & Radford, L. (2011). *A cultural-historical perspective on mathematics teaching and learning*. Rotterdam: Sense.
Roth, W.-M., & Thom, J. (2008). Bodily experience and mathematical conceptions: From classical views to a phenomenological reconceptualisation. *Educational Studies in Mathematics, 70*, 175–189.
Roth, W.-M., & Walshaw, M. (2019). Affect and emotions in mathematics education: Toward a holistic psychology of mathematics education. *Educational Studies in Mathematics, 102*(1), 111–125.
Rowland, T. (2012). Contrasting knowledge for elementary and secondary mathematics teaching. *For the Learning of Mathematics, 32*(1), 16–21.
Rowland, T., & Ruthven, K. (Eds.). (2011). *Mathematical knowledge in teaching*. Dordrecht: Springer.
Saenz-Ludlow, A., & Presmeg, N. (2006). Guest editorial. Semiotic perspectives on learning mathematics and communicating mathematically. *Educational Studies in Mathematics, 61*(1–2), 1–10.
Sant, E., & Brown, T. (2020). The fantasy of the populist disease and the educational cure. *British Educational Research Journal*.

Schürmann, U. (2019). The separation of mathematics from reality in scientific and educational discourse. In P. Ernest (Ed.), *The philosophy of mathematics education today* (pp. 241–250). New York: Springer.

Schütz, A. (1962). *The problem of social reality*. The Hague: Martinus Nijhof.

Schwab, J. (1983). The practical 4: Something for curriculum professors to do. *Curriculum Inquiry, 13*(3), 239–265.

Shulman, L. (1986). Those who understand: Knowledge growth in teaching. *Educational Researcher, 15*(2), 4–14.

Skemp, R. (1976). Relational understanding and instrumental understanding. *Mathematics Teaching, 77*, 20–26.

Skott, J. (2014). Towards a participatory approach to beliefs in mathematics education. In B. Pepin & B. Roesken-Winter (Eds.), *From beliefs to dynamic affect systems in mathematics education*. Dordrecht, Springer.

Skovsmose, O. (1994). *Towards a philosophy of critical mathematics education*. Dordrecht: Kluwer.

Skovsmose, O. (2016). An intentionality interpretation of meaning in mathematics education. *Educational Studies in Mathematics, 92*(3), 411–424.

Skovsmose, O. (2019). Students' foregrounds and politics of meaning in mathematics education. In P. Ernest (Ed.), *The philosophy of mathematics education today* (pp. 115–130). New York: Springer.

Smith, P. (1988). *Discerning the subject*. Minneapolis: University of Minnesota Press.

Smith, K., Hodson, E., & Brown, T. (2013). The discursive production of classroom mathematics. *Mathematics Education Research Journal, 25*, 379–397.

Solomon, Y. (1998). Teaching mathematics: Ritual, principle and practice. *Journal of Philosophy of Education, 32*(3), 377–390.

Solomon, Y. (2008). *Mathematical literacy: Developing identities of inclusion*. London: Routledge.

Spanjaard, M., Reeves, N., van Dieen, J., Baltzopoulos, V., & Maganaris, C. (2008). Lower-limb biomechanics during stair descent: Influence of step-height and body mass. *Journal of Experimental Biology, 211*, 1368–1375.

Sriraman, B., & English, L. (2010). *Theories of mathematics education: Seeking new frontiers*. Springer.

Steffe, L., & Kieran, T. (1994). Radical constructivism and mathematics education. *Journal for Research in Mathematics Education, 25*(6), 711–733.

Strathern, M. (2000). *Audit cultures: Anthropological studies in accountability, ethics and the academy*. London: Routledge.

Strittmatter, K. (2019). *We have been harmonised: Life in China's surveillance state*. London: Old Street.

Tang, S., Morgan, C., & Sfard, A. (2012). Investigating the evolution of school mathematics through the lens of examinations: developing an analytical framework. Paper presented at the 12th *International Congress on Mathematical Education*, Seoul, Korea.

Tessier, B. (2012). Mathematics and narrative: Why are stories and proofs interesting? In A. Doxiadis & B. Mazur (Eds.), *Circles disturbed: The interplay of mathematics and narrative* (pp. 232–243). Princeton: Princeton University Press.

Thiel, J. (2018a). *The production of student voice and its effects on academia*. PhD thesis, Manchester Metropolitan University.

Thiel, J. (2018b) Student feedback apparatuses in higher education: an agential realist analysis. *Discourse: Studies in the Cultural Politics of Education.*

Thiel, J. (2019). The UK National Student Survey: An amalgam of discipline and neo-liberal governmentality. *British Educational Research Journal, 45*(3), 538–553.

Tomšič, S., & Zevnik, A. (2016). *Jacques Lacan; Between psychoanalysis and politics*. London: Routledge.

von Glasersfeld, E. (Ed.). (1991). *Radical constructivism in mathematics education*. Dordrecht: Kluwer.

von Glasersfeld, E. (1995). *Radical constructivism: A way of knowing and learning*. London: Falmer.

Wagner, D. (2012). Opening mathematics texts: Resisting the seduction. *Educational Studies in Mathematics, 80*(1–2), 153–169.

Walkerdine, V. (1988). *The mastery of reason*. London: Routledge.

Walls, F. (2009). *Mathematical subjects: Children talk about their mathematical lives*. New York: Springer.

Walshaw, M. (Ed.). (2004). *Mathematics education within the postmodern*. Greenwich, CT: Information Age Publishers.

Walshaw, M. (2008). Developing theory to explain learning to teach. In T. Brown (Ed.), *The psychology of mathematics education*. Rotterdam: Sense Publishers.

Walshaw, M. (Ed.). (2010). *Unpacking pedagogy: New perspectives for mathematics classrooms*. Charlotte, NC: Information Age Publishing.

Walshaw, M., & Brown, T. (2012). Affective productions of mathematical experience. *Educational Studies in Mathematics, 80*(1–2), 185–200.

Watson, A. (2017). Pedagogy of variations: Synthesis of various notions of various pedagogy. In R. Huang & Y. Li (Eds.), *Teaching and learning mathematics through variation. Mathematics teaching and learning*. Rotterdam: Sense Publishers.

Watson, A., & Winbourne, P. (Eds.). (2008). *New directions for situated cognition in mathematics*. Dordrecht: Springer.

Wenger, E. (2000). Communities of practice: Learning, meaning and identity. Cambridge: Cambridge University Press.

White Rose Maths. (2020). *Secondary Resources*. [Online] [accessed on 22 May 2020] https://whiterosemaths.com/resources/secondary-resources/

Williams, R. (1983). *Keywords*. London: Flamingo.

Williams, P. (2008). *Independent review of mathematics teaching in early years settings and primary schools*. London: Department for Children, Schools and Families.

Williams, J. (2012). Use and exchange value in mathematics education research: Contemporary CHAT meets Bourdieu's sociology. *Educational Studies in Mathematics, 80*(1–2).

Williams, J. (2015). Alienation in mathematics education: Critique and development of neo-Vygotskian perspectives. *Educational Studies in Mathematics, 92*(1), 59–73. *(early draft)*.

Williams, J. (2019). Mastery mathematics but who is the slave? *Education for tomorrow*, 1. [Online] [Accessed 18 Jan 2020] https://educationfortomorrow.org.uk/mastery-mathematics-but-who-is-the-slave/

Wittgenstein, L. (1983). *Philosophical investigations*. Oxford: Basil Blackwell.

Zan, R., Brown, L., Evans, J., & Hannula, M. (2006). Affect in mathematics education: An introduction. *Educational Studies in Mathematics*.

Zeichner, K. (2010). Competition, economic rationalization, increased surveillance, and attacks on diversity: Neo-liberalism and the transformation of teacher education in the US. *Teaching and Teacher Education, 26*, 1544–1552.

Žižek, S. (1989). *The sublime object of ideology*. London: Verso.

Žižek, S. (1997). *The plague of fantasies*. London: Verso.

Žižek, S. (1998). Four discourses. Four subjects. In S. Žižek (Ed.), *Cogito and the unconscious*. Durham, NC: Duke University.

Žižek, S. (2001). *Did somebody say totalitarianism?* London: Verso.

Žižek, S. (2006a). *How to read Lacan*. London: Granta.

Žižek, S. (2006b). *The parallax view*. London: MIT press.

Žižek, S. (2011). The spectre of ideology. In I. Szeman & T. Kaposy (Eds.), *Cultural theory: An anthology*. Chichester: Wiley-Blackwell.

Žižek, S. (2012). *Less than nothing: Hegel and the shadow of dialectical materialism*. London: Verso.

Žižek, S. (2014). *Event*. London: Penguin.

Žižek, S. (2017). *The incontinence of the void: Economico-philosophical spandrels*. London: MIT Press.

Žižek, S. (2020). *Sex and the failed absolute*. London: Bloomsbury.

Žižek, S., & Millbank, J. (2009). In C. Davis (Ed.), *The monstrosity of Christ: Paradox or dialectic?* Cambridge, MA, The MIT Press.

Index

T. Brown, *A Contemporary Theory of Mathematics Education Research*, https://doi.org/10.1007/978-3-030-55100-1

- Written in an easy conversational style (emoji p19)

- pp 22 Mathematical entities are built in, and then onto, human self-image.
  (We build models of exam success and then identify ourselves in terms of them).

- The structure, drawing on past work in each chapter, means that there is some repetition of language/ideas.
  Also find repeated sentences which one assumes are from editing previous material eg p 29/33 + p 40; poor language p44.

- Different theoretical frames demand a lot of the reader – e.g. Ch4 which introduces a Lacanian frame and uses it – but not enough detail for me to follow fully. Felt like the analysis was rather superficial ... PhD > Article > Chapter!
- And then Žižek is embedded throughout and might have been more fully articulated??

- Good to have a book which focuses on the relationship between society and knowledge.

  → Rather theoretically hotch-potch?

- Ch5 – Student teachers' M conceptions, was weak IMO.
  Not clear what it was trying to do/say; poorly theorised – felt like it needed Bernstein or similar.

Ch 6 - I couldn't really see what this was about

Ch 7 + generally — falls into the trap of trying to weave complex theoretical ideas into the text as needed for the argument .. but they remained unintelligible to me!

Whole chapter is unintelligible - indeed I challenge anyone to be able to make sense of it. Seems simply to be arguing that trainees need to learn to see how maths is constructed and to remain open to not understanding.

Complete cobblers!!

Ch 8 "Newer theories have the disadvantage of needing to express themselves through familiar pathways..." (150)

p121 - Isolation of mathematics from social context! YES!

Meeting 17/1/22.

Lacan – subject of desire – always looking to close the gap between the story and the real.

Essentially arguing that V. & P. normalise M. Research.

Lightning Source UK Ltd.
Milton Keynes UK
UKHW021040161121
394057UK00001B/30

9 783030 550998